# An Outline of
# Informational Genetics

# Synthesis Lectures on Biomedical Engineering

**Editor**
John D. Enderle, *University of Connecticut*

An Outline of Informational Genetics

Gérard Battail


ISBN: 978-3-031-00501-5        paperback
ISBN: 978-3-031-01629-5        ebook


DOI 10.1007/978-3-031-01629-5


*A Publication in the Springer series*
*SYNTHESIS LECTURES ON BIOMEDICAL ENGINEERING*


Lecture #23
Series Editor: John D. Enderle, University of Connecticut

Series ISSN
Synthesis Lectures on Biomedical Engineering
ISSN pending.      Print 1930-0328   Electronic 1930-0336

# An Outline of Informational Genetics

Gérard Battail

École Nationale Supérieure des Télécommunications (retired), Paris

*SYNTHESIS LECTURES ON BIOMEDICAL ENGINEERING #23*

# ABSTRACT

Heredity performs literal communication of immensely long genomes through immensely long time intervals. Genomes nevertheless incur sporadic errors referred to as mutations which have significant and often dramatic effects, after a time interval as short as a human life. How can faithfulness at a very large timescale and unfaithfulness at a very short one be conciliated?

The engineering problem of literal communication has been completely solved during the second half of the XX-th century. Originating in 1948 from Claude Shannon's seminal work, information theory provided means to measure information quantities and proved that *reliable* communication is possible through an *unreliable* channel (by means left unspecified) up to a sharp limit referred to as its capacity, beyond which communication becomes impossible. The quest for engineering means of reliable communication, named *error-correcting codes*, did not succeed in closely approaching capacity until 1993 when Claude Berrou and Alain Glavieux invented turbocodes. By now, the electronic devices which invaded our daily lives (e.g., CD, DVD, mobile phone, digital television) could not work without highly efficient error-correcting codes. Reliable communication through unreliable channels up to the limit of what is theoretically possible has become a practical reality: an outstanding achievement, however little publicized.

As an engineering problem that nature solved aeons ago, heredity is relevant to information theory. The capacity of DNA is easily shown to vanish exponentially fast, which entails that error-correcting codes must be used to regenerate genomes so as to faithfully transmit the hereditary message. Moreover, assuming that such codes exist explains basic and conspicuous features of the living world, e.g., the existence of discrete species and their hierarchical taxonomy, the necessity of successive generations and even the trend of evolution towards increasingly complex beings.

Providing geneticists with an introduction to information theory and error-correcting codes as necessary tools of hereditary communication is the primary goal of this book. Some biological consequences of their use are also discussed, and guesses about hypothesized genomic codes are presented. Another goal is prompting communication engineers to get interested in genetics and biology, thereby broadening their horizon far beyond the technological field, and learning from the most outstanding engineer: Nature.

# KEYWORDS

error-correcting codes, information theory, molecular genetics, nested codes, soft codes.

# Contents

# Foreword

The scientific and engineering literature uses an impersonal style, where the author tradition-ally refers to himself or herself as *we*. Science and engineering strive for objectivity, so subjectivity should be avoided in this literature as somewhat indecorous. I depart from this tradition in this Foreword but also in Part I, entitled "An Informal Overview," because its content depends to a large extent on my personal experience as a communication engineer. Moreover, an informal style is perhaps the best way to introduce this lecture so as to motivate its reading.

Until my retirement in 1997, I taught information theory and error-correcting codes for almost 25 years at the *École nationale supérieure des Télécommunications* in Paris, France. Then, as a hobby, I contemplated applications of these topics to sciences of nature. The present lecture is devoted to expounding some results of this research as regards genetics.

Molecular genetics is based on the postulate that DNA is a long-term memory which faithfully conserves the genetic information during time intervals at the geological scale. The trouble is that DNA is subject to sporadic errors, or mutations, which have dramatic effects on the living beings, like aging or cancer, at the much shorter timescale of individual lifetimes. The existence of such mutations blatantly contradicts the assumed long-term conservation of the genetic message borne by DNA. Indeed, DNA may be interpreted as a channel for communicating through time. Information theory associates with any channel a fundamental quantity which measures its ability to convey information, its *capacity* . The capacity actually sets an impassable limit for any communication using the channel. One easily computes the DNA capacity as a function of time, showing beyond any doubt that it vanishes as time increases unless the mutation rate is exactly zero. The mechanism alleged by geneticists, i.e., copying it using the Watson-Crick template-replication mechanism, thus fails to account for the faithful communication of the genetic message through time.

On the other hand, communication engineers developed very powerful tools, referred to as error-correcting codes, to cope with errors occurring in communication channels. They are by now widely used in objects of daily life like computer memories, mobile phones, CD, DVD, digital television, etc. They enable *reliable* communication through *unreliable* channels, what information theory proves to be possible however paradoxical it may look (and the engineering practice plainly confirms). Assuming that such tools are actually implemented in genomes is the only way to account for the faithful conservation of the genetic message. Extremely inventive, nature has widely predated and often surpassed human engineers, which makes this necessary assumption highly plausible. Most geneticists are not aware of the extreme difficulty of conserving genomes during geological times and ignore that communication engineers know how it can be performed. Information theory shows that error-correcting codes are actually the only possible solution to this crucial but overlooked biological problem. It turns out, unfortunately, that both the theoretical possibility of 'error-free' communication and the practical means to implement it are almost unknown, except of course by communication engineers.

Let me illustrate the situation with a fable. Imagine a world where physicists and biologists are strictly separated. The lack of communication between both communities would possibly result in biologists explaining facts of the living world by means that physicists know to be impossible. For instance, imagine that biologists explain the flight of flies, bats, and birds by assuming that wings cancel gravity (anti-gravity is an old trick of science fiction). Establishing a communication between both communities would sooner or later ruin the biologists' explanation of animal flight. Now replace physicists by information-theorists, experts in animal flight by molecular geneticists, and their assumed separation by the ever-increasing specialization of scientific disciplines and their recourse to technical jargons. The situation is comparable and even worse. First of all, nobody ever proved that anti-gravity is physically impossible, while the channel capacity of information theory is a mathematically proven impassable limit. Second, and more importantly, animal flight is not such a central point while the conservation of genomes is the cornerstone of biology as a whole: indeed, the very existence of the living world crucially depends on it. Fortunately, however, information theory does not only prove that merely copying the genomes cannot ensure their conservation but also tells what means can perform it, namely, error-correcting codes.

I do not mean that biologists failed to realize that information is a crucial component of life phenomena. Indeed, many biologists clearly understood its importance, especially in heredity. Maynard Smith quotes Weissmann as having first realized it was so. For instance, he wrote [58]: '[Weissmann] understands that heredity is about the transmission of information, and not just matter and energy.' Although Maynard Smith endorsed this opinion, he quickly discarded information theory, arguing that it could be of no use in biology because it ignores semantics [60, 73]. This is a fundamental misunderstanding. Ignoring semantics is not a weakness of information theory, but maybe its strongest methodological position. As far as literal communication is concerned (as performed by telecommunication means), semantics is indeed not relevant: the meaning of a message does no affect its transportation, and a messenger who carries it has not to know about its content. Rejecting information theory for this reason is just *lâcher la proie pour l'ombre* (literally, letting go the prey for the shadow), i.e., giving up a coherent, complete and quantitative science of communication for no science at all. Despite the efforts of Yockey who pleaded for the use of information theory in biology [79, 80], the position of Maynard Smith has been widely adopted by biologists. (As requiring no educational effort, it is also the laziest one.) Then 'biological information' became a fuzzy, undefined concept except that, in a purely negative way, it was claimed to differ from engineering information. The word 'information' has become ubiquitous in the biological literature but most often used with a loose meaning. In my opinion, *a priori* rejecting information theory is as poorly justified and as harmful to the progress of biology as has been in its time the now obsolete concept of an irreducible 'organic chemistry.'

This lecture is intended for both geneticists and communication engineers. The impossibility of achieving faithful genetic transmission by mere copying is easily understood by the latter who know that the channel capacity is a strictly impassable limit and who can easily check that the capacity of DNA, considered as a channel for communicating through time, is actually a vanishing function of time. However, a vast majority of geneticists have no knowledge at all of information theory so an argument based on the decrease of DNA capacity is mere dead letter for them. Before I can convince them that mere copying is not the solution but that error-correcting codes are necessarily involved, I have to discuss information theory up to the point where they fully understand the meaning of capacity and the operation of error-correcting codes. A large part of this lecture must thus be devoted to this end. Most geneticists would have thrown out the book well before this point! Although communication engineers will accept my argument, they will probably feel uncomfortable with molecular genetics, so I must present some sketchy discussion of this topic which, as always in biological matters, is quite complex in its details. Intending this lecture for both geneticists and communication engineers, I must thus spend many pages discussing matters which reveal to be strongly connected to each other only at the end of the book.

This is why Part I is a summary of the lecture content in an informal style serving as a common ground for both biologists and engineers, hopefully overcoming their different cultural habits and methods. Much like the characters of a novel are progressively revealed as the reader proceeds, it does not comply with the rule of defining all the scientific words and concepts at their first appearance. Part II brings the precision and details which Part I lacks. Part III states, in the usual scientific style, the hypotheses that information theory shows to be necessary in order to explain the genome conservation, and discusses their consequences on the living world and its evolution. Many of their actual features find here an explanation, and answers are proposed to debated questions. I refer to the mandatory conditions stated above as hypotheses, although the word 'hypothesis' is too weak and has a speculative connotation, because I am yet unable to exhibit the details of genomic error-correcting codes. Furthermore, the means which implement them and achieve genome conservation remain to be identified. Making them fully explicit is a formidable task which exceeds by far the means available to me. It requires a collaboration of communication engineers and practicing geneticists which remains to be set up.

A collaboration of engineers and geneticists needs of course a common language. Engineers developed information theory as the science of communication engineering. As a mathematical science, it needed a precisely defined vocabulary. Biologists then borrowed this vocabulary to communication engineering, but used it loosely, in a descriptive rather than mathematical manner. For instance, the word 'code' belongs to the vocabulary of information theory since its very beginning, i.e., Shannon's paper [72], but geneticists began using the phrase 'genetic code' in the 1960s. What 'code' means for them is rather different from the information-theoretic meaning of the word, of which they were apparently unaware. I'll thus need to be very fussy in regards to definitions and vocabulary appropriateness. This runs the risk of looking pedantic, but it is the price to be paid for avoiding ambiguity.

Again for avoiding ambiguity, I'll name 'shannon' the binary unity of information quantity, at variance with the current information-theoretic literature which, following Shannon himself [72], names it 'bit', an acronym for 'binary digit.' However, a binary digit does not always bear information, and when it does the information quantity it bears is not always this unity. A unit and a digit are moreover definitely different entities, so using different words for naming them is better. Naming 'shannon' the unit of information quantity moreover made me free to use 'bit' as a convenient abbreviation for 'binary digit', especially in Chs. 5 and 6.

No special effort has been made to avoid redundancy. Some of the main ideas are developed in different chapters of this lecture, hopefully shedding more light on the subject when dealt with in different perspectives. Just like redundancy is the key for eliminating errors in communication engineering, I believe it is necessary in the semantic field for an efficient didactic communication.

---

Attempts were made earlier to apply information theory to biology. Undeniably, the great pioneer in this field is Yockey [79, 80]. Besides their scientific content, his books castigate the 'apartheid in the olive grove of academe' which prevents a majority of geneticists to realize that molecular biology actually needs information theory. I fully concur with this criticism. Not only do biologists ignore potentially relevant and useful tools, but it is a pity that ideas that information theory shows to be wrong are still taken as true by many of them and taught in textbooks.

As far as I know, the first attempt to identify error-correcting codes in genetics was made in 1981 by Forsdyke [34], who later gathered in *Evolutionary Bioinformatics* [37] the fruit of his rich experience. Many topics dealt with in this excellent book illustrate, in biological rather than engineering terms, the concept of 'soft code' I discuss in this lecture (Ch. 10). I read it too late to take its content into account here, but I'll hopefully exploit it in further works as will researchers wishing to work along the same line as me.

I would like to also mention that the book by Barbieri, *The Organic Codes* [5], proposes a picture of the living world rather similar to mine through a very different approach. The 'codes' of the title are not error-correcting codes *stricto sensu*, but they may well be interpreted in terms of 'soft codes,' a concept to be discussed in Ch. 10 below. Moreover, they assume the structure of 'nested codes,' another concept I think is necessary for explaining the permanency of the older information, which is discussed in Ch. 9.

Thanks are due to Daniel Mange of the *Ecole polytechnique fédérale de Lausanne* in Switzerland who gave me the opportunity to meet Marcello Barbieri of the University of Ferrara, Italy. I found in him an enthusiastic and friendly interlocutor who generously invited me several times to discuss my ideas both orally and in writing; I am of course deeply indebted to him. I also sincerely thank other people who manifested interest in my work and gave me opportunities to discuss the ideas contained in this lecture or with whom I exchanged ideas, especially Claude Berrou, John Enderle, Paddy Farrell, Donald R. Forsdyke, Jean-Michel Labouygues, Vittorio Luzzati, Elebeoba E. May, Olgica Milenkovic, Mark E. Samuels, and Hubert P. Yockey.

Gérard Battail
École Nationale Supérieure des Télécommunications (retired), Paris
September 2008

# Part I

# An Informal Overview

CHAPTER 1

# Introduction

*If you want to understand life, don't think about vibrant, throbbing gels and oozes, think about information technology.*

Richard Dawkins, *The Blind Watchmaker*

## 1.1 GENETICS AND COMMUNICATION ENGINEERING

Although at first sight they may look unrelated, genetics and communication technology are both concerned with *transfer of information*. Besides the obvious fact that the latter is a matter of human engineering while the former is concerned with natural processes, they mainly differ in not operating in the same dimension: the latter is intended to communicate messages in space, from one place to another, while the former communicates hereditary messages in time, from one instant to another.

The outstanding success of communication technology relies of course on tremendous progresses in physical hardware but also, much less visibly, on the development of very powerful conceptual tools, consistently gathered in *information theory* [72]. Although these tools were originally intended for communication through space, they are comprehensive enough to apply as well to communication through time. The central question to which this lecture proposes an answer is:

*Can the theoretical framework elaborated by human engineers shed light on the natural processes of hereditary communication?*

The answer is yes, beyond any expectation: the perspective provided by information theory deeply renews the vision we can have of the living world. It definitely shows that the template-replication paradigm now in force does not account for the faithful conservation of genomes. This negative result would suffice to call for a deep reappraisal of the fundamentals of genetics, hence of biology as a whole.

I can thus claim that conventional biology gives an inadequate answer to the crucial question of genome conservation. Since the discovery of its double-helix structure, the replication of DNA using one of its strands as a template has become the paradigm of genome conservation and it has remained almost unquestioned. The starting point of the research reported in this lecture is its questioning, from an outer perspective, that of communication engineering. As defined by information theory, the capacity of a channel is the quantity which measures its ability to convey information. Considering DNA as a channel for communicating through time, it is a trivial matter to compute its capacity or, more precisely, a simple upper bound on it. It turns out that it vanishes exponentially fast due to the cumulative effect of casual errors. The memory of heredity—DNA—is thus not permanent as believed but actually ephemeral at the geological timescale. Without further means to ensure the permanence of the genomic message, DNA replication cannot account for its faithful communication. Genome conservation is not the rule and error is not the exception.

On the positive side, information theory asserts that faithful genome conservation can actually be secured provided very efficient *error-correcting codes* are used. One *must* thus assume that it is so. This assumption will be referred to as the *main hypothesis*. (I already explained in the Foreword why I use the word 'hypothesis' for referring to a mandatory condition, despite its weakness and speculative connotation.) Accounting for the faithful conservation of very old information leads, furthermore, to assume as a subsidiary hypothesis that the genomic error-correcting code takes the form of a layered structure—referred to as 'nested codes' or 'nested system'—which combines several codes which successively appeared in time: new information has been appended to a previously encoded message, resulting in a sequence which has been further encoded, and this process has been repeated several times in the past. Information is thus encoded by a number of component codes, the larger, the older it is. Being encoded by more numerous component codes, old information is better protected against errors than the more recent.

Although they properly recognized the importance of permanence in the evolutive success of a species, it seems that biologists did not realize how difficult it is to maintain the information integrity through the ages. Dawkins wrote about the needed accuracy of replication [29, pp. 16–17]: 'We do not know how accurately the original replicator molecules made their copies. Their modern descendants, the DNA molecules, are astonishingly faithful compared with the most high-fidelity human copying process […]'. Dawkins expresses his astonishment as regards such a high reliability but he *does not question about the means to obtain it* nor the consequences which may result from their use. It will be shown below that high reliability cannot actually result from mere copying, but that intrinsic properties of the genome must make it resilient to casual errors.

For ensuring the genome permanence, the template-replication of DNA actually needs to be complemented with error-correcting means. Genomic messages must be endowed with error-correcting properties, and their conservation can only result from a dynamic process involving these intrinsic properties. Moreover, the genome regeneration based on them must be performed after a short enough time interval, in order to avoid the channel capacity becoming too small or, in terms of error correction, that the cumulated number of errors in the genome exceeds its intrinsic error-correcting ability.

Once it is clear that the conservation of genomes can result from a dynamic process which succeeds only when stringent conditions are met, one may expect these conditions to have a great impact on the living world as a whole, due to the central role that genomes play in it. Indeed, although no explicit description of the assumed genomic code can yet be given, the main and subsidiary hypotheses just stated provide explanations to many basic features of the living world yet unexplained, some of them because biologists did not think they need an explanation and preferred to deal with them as basic postulates. These hypotheses also hint at answers to some debated questions. Moreover, no irreducible contradictions between them and established biological facts were found although, of course, some points will need further elaboration.

The ideas to be developed below have their seeds in a number of papers and symposium presentations, among which [13, 15, 16, 17, 18, 19].

## 1.2   SEEING HEREDITY AS A COMMUNICATION PROCESS EXPLAINS FEATURES OF THE LIVING WORLD

### 1.2.1   MAIN AND SUBSIDIARY HYPOTHESES

For emphasis, we state again here the two 'hypotheses' on which this lecture is based. Most of Part III will be devoted to develop the consequences of these hypotheses and to compare them with features of the actual living world.

*Main hypothesis:*   Any genome actually belongs to an efficient error-correcting code, to be referred to below as *genomic code*. Its necessity stems from the need of a faithful conservation of the genetic information.

*Subsidiary hypothesis:*   The genomic error-correcting code postulated by the main hypothesis actually takes the form of a layered structure, to be referred to as *nested system*, which combines several codes which successively appeared in time. This hypothesis is actually needed in order to account for the faithful conservation of very old genetic information.

Chapters 3 and 6 deal with error-correcting codes. Precisely describing the nested system will be deferred to Ch. 9. Although several codes are involved in the nested system postulated by the subsidiary hypothesis, the phrase 'genomic code' will be used to collectively designate the set of codes which, combined into the nested system, contribute to making the genome resilient to errors. An error-correcting code will often be restricted to sequences which satisfy some set of constraints. Then the word 'code' will be used in two different but related meanings: to designate either this set of sequences, or the set of constraints they satisfy. Doing so hopefully avoids lengthy periphrases, the context being most often sufficient to avoid any ambiguity.

Anticipating on what follows, the picture of the living world which derives from the above hypotheses is now roughly outlined.

### 1.2.2   A STATIC VIEW OF THE LIVING WORLD: SPECIES AND TAXONOMY

An error-correcting code is a collection of symbol sequences, referred to as its *words*, which are as different as possible from each other. A very convenient geometrical language, which moreover has a rigorous mathematical meaning, interprets these codewords as points of an abstract space having as many dimensions as the words have symbols. Each coordinate of these points is associated with a symbol at a precise location in the words and assumes numerical values which unambiguously represent the alphabet symbols. The distance between two points in this space is moreover defined as the number of locations where their coordinates differ (i.e., the number of symbols where the corresponding words differ). It is referred to as the *Hamming distance* and will be used throughout in the rest of the book. The messages to be communicated are assumed to belong to a given finite set. A set of words, as many as the possible messages, is used for communicating them. It is referred to as a *code*. Each word of this code represents one of the messages according to a one-to-one correspondence. Communicating a message consists of transmitting the codeword which corresponds to it. The symbol errors which affect the received word tend to diminish its distance with respect to other codewords but, if the errors are not too many, the received word remains closer to the actually transmitted word than to any other codeword. Then the actually transmitted word can be unambiguously determined. To be efficient, an error-correcting code should thus have its words as distant as possible from each other. The smallest distance between any two different words of an error-correcting code is referred to as its *minimum distance*.

In genetics, the symbols are *nucleic bases*, also referred to as *nucleotides*, the succession of which along the DNA molecule constitutes the genomic message. Assuming that a genomic error-correcting code exists implies that genomes are neatly separated, hence the very existence of discrete species is a mere consequence of the main hypothesis. Moreover, the subsidiary hypothesis implies a distance hierarchy within the layers of the nested system: the words which belong to the central (older) layers are more distant from each other than those of the peripheral (more recent) ones. Not only the biological fact that distinct species exist is explained, but also the very possibility of their ordering according to a hierarchical taxonomy.

It has been stated already that the genome regeneration based on its error-correcting properties must be performed after a short enough time interval. That life proceeds by successive generations is a trivial biological fact, but it receives here an explanation which greatly enhances its significance. 'Successive generations' must now be understood as 'successive *regenerations*,' meaning that each new living being starts with an exactly regenerated genome, except for the very few ones which suffered a regeneration error.

### 1.2.3   A DYNAMIC VIEW OF THE LIVING WORLD: EVOLUTION

An error-correcting code endows a genome with resilience to casual errors, hence making its exact recovery possible despite the errors which may affect the nucleotides. It works only within certain limits beyond which another word, at a distance of at least the minimum distance of the code, results from attempting to recover the transmitted word. This very unusual event transforms a genome into a neatly different one, hence originating in a new species. Let us now elaborate on the dynamic view of the living world which results from this remark.

The engineering function of decoding aims at recovering the codeword which has been transmitted in order to represent some message, given the corresponding received sequence of symbols, i.e., the transmitted codeword affected by errors. Its first step is intended to identify the actually transmitted codeword, and in the genomic context it will be referred to as *regeneration*. The second step, i.e., recovering the encoded message, is trivial once the first one is performed since a one-to-one correspondence between the possible messages and the codewords has been assumed. Regeneration is made possible because distances separate the codewords. The optimum regeneration rule consists of choosing the word of the code the closest to the received word. If the symbol errors are too many, it may occur that the word chosen according to this rule is not the correct one and, in the presence of casual errors, such a regeneration error occurs with nonzero probability. Contrary to intuition, the probability of a regeneration error can be made arbitrarily small by *increasing* the length of the codewords (also referred to as the length of the code) and using a properly designed code, but only provided the average quantity of information borne by each symbol of the codeword is smaller than a fundamental quantity, referred to as the *capacity* of the channel, which depends on the symbol error probability. If this inequality is not satisfied, a regeneration error occurs almost certainly, whatever the code and means intended for regenerating the transmitted codeword.

In normal conditions, the time interval between regenerations is as short as to make the probability of regeneration failure extremely small. It is necessarily so to account for the stability of species which we observe at the scale of individual lifetimes. As an extremely infrequent event which changes a genome into another neatly different one, we may think of a regeneration error as the possible origin of a new species. The main hypothesis thus simply leads to a saltationist vision of biological evolution. Moreover, the subsidiary hypothesis hints at a phyletic tree which coincides with the taxonomic hierarchy since the higher in this hierarchy is a component code, the larger its minimum distance, hence the more infrequent is a regeneration error at its hierarchical level.

Assuming that species originate in regeneration errors also suggests their contingency, since regeneration errors are chance events. It should, however, not be concluded that the genomic messages are purely random: on the contrary, they actually obey the constraints which define the genomic error-correcting code. The wrong recovery of a codeword is a chance event which concerns the word *as a whole*; each of its symbols is not chosen at random. This remark is important since it refutes arguments from the 'intelligent design' upholders on the improbability of errors simultaneously affecting two different nucleotides. The probability of two simultaneous errors equals the square of the probability of error in a single symbol (and thus is much smaller) only if the symbol errors are independent events. Since the error event consists of choosing a codeword as a whole, the symbol errors at two specified locations are no longer independent events so the argument does not hold.

Species do not only originate in regeneration errors. Transpositions, chromosome rearrangements, and integration of genetic material of outer origin, especially viral, are already known mechanisms which deeply modify genomes. The setting up of a new code in the nested system assumed according to the subsidiary hypothesis necessarily results from an event of this kind. A better knowledge of the genomic error-correcting codes will be necessary in order to understand their possible connection with such phenomena.

The main hypothesis suffices to explain the yet poorly understood trend of evolution towards increasing *complexity*. It has been stated above that the probability of a regeneration error can be made arbitrarily small by increasing the length of the codewords, although the longer the codewords, the larger the average number of errors which affect them. The ability to survive, thanks to a low probability of regeneration errors, is an evolutive advantage, just as the ability of fast reproduction. The ability of fast reproduction is limited by the genome length, which should be as large as to specify the replication machinery (if we except viruses which use the cellular machinery of their hosts for their own reproduction), so there is a lower limit to this length. On the other hand, there is no limit to the evolutive advantage which can result from the lengthening of genomes: the average genome lifetime which results from using an error-correcting code varies as the reciprocal of the probability of a regeneration error, as shown in Ch. 8. Lengthening the genome can result in an arbitrarily small regeneration error probability, hence there is no limit to the increase in average lifetime it can provide. The subsidiary hypothesis of a nested system moreover implies that lengthening the genome by introducing one more coding layer enhances the code minimum distance, hence diminishes the regeneration error probability of the already present genomic information.

## 1.3   REGENERATION VERSUS REPLICATION

Replication and regeneration are deeply different functions. Their confusion would make this lecture incomprehensible. Let us summarize their difference before we proceed further.

Replication makes a copy of the original genome as faithful as possible, thus ideally identical to it. Since copying an already erroneous genome just keeps the errors it bears, it does not faithfully ensure the long-term conservation of genomes.

Regeneration is intended to rewrite the genomic message in such a way that:
— the rewritten message strictly satisfies the constraints of the genomic code; and
— is the closest to the received genomic message possibly affected by errors.

Then, the casual errors which may affect the genome are corrected, except if the number of these errors exceeds the error-correcting ability of the code. For a properly designed code and a short enough interval between regenerations, the probability of a regeneration error is extremely small. The crucial importance of conserving genomes necessarily led natural selection to a good enough code and a short enough time interval. Regeneration in itself does not suffice to ensure the survival of a species which needs that the number of its members increases, so a regenerated genome must be replicated.

Notice that, contrary to replication, regeneration does not result in a message faithful to the original one. However, it is faithful to the genomic code seen as a set of constraints. The genomic code appears here as more important than the individual genomes the conservation of which it ensures. Implementing regeneration, unlike replication, is costly in terms of processing complexity.

The above topics will be dealt with at greater length in the following. Before I can proceed, I must roughly describe molecular genetics, and briefly expound information theory and error-correcting codes. These overviews are somewhat lengthy but I have been unable to make them shorter without hampering their intelligibility.

# CHAPTER 2

# A Brief Overview of Molecular Genetics

## 2.1 DNA STRUCTURE AND REPLICATION

Excellent books popularize molecular genetics, among which I particularly appreciated those by Dawkins [29, 30] for my own initiation to the topic (my background is in engineering). They were especially helpful to me as they provided a neatly defined model.

Everybody by now knows that the agent of genetic communication is DeoxyriboNucleic Acid (DNA). It is a very long unidimensional polymer which assumes the form of a double helix. Each strand of the double helix consists of a regular succession of deoxyribose molecules alternating with phosphate groups. Deoxyribose, a sugar, is tied to the phosphate groups by covalent bonds. The deoxyribose–phosphate chain has a mechanical role but does not convey information (it is often metaphorically referred to as the DNA 'backbone'). One of the four 'nucleic bases' (or 'nucleotides')— adenine (**A**), thymine (**T**), guanine (**G**), and cytosine (**C**)—is covalently bound to each sugar molecule of the deoxyribose–phosphate chain. The succession of nucleotides according to a well-defined direction along the backbone of one strand constitutes a message written using the four-letter alphabet {**A, T, G, C**}. The genetic information is borne by this message where the nucleotides can *a priori* assume any order, so a string of $n$ nucleotides can bear in principle $4^n$ different messages. The other strand in the double helix structure bears a similar message which is entirely determined once the message borne by the first strand is given. In front of each nucleotide **A** of one strand, **T** is found on the other one; and **C** in front of **G**. In other words, the nucleotides appear in *complementary pairs*, **A–T** and **G–C**. The DNA molecule then appears as a (twisted) ladder, with the molecular 'backbones' as uprights and the pairs of complementary nucleotides as rungs. DNA as described is the material support of heredity for all living things (if we except some viruses where this function is assumed by ribonucleic acid, RNA, another unidimensional polymer whose chemical structure and properties are similar to DNA). DNA is combined with proteins in large structures named 'chromosomes.' In prokaryotes, i.e., unicellular beings having no nucleus, the total DNA content is referred to as its 'genome'. In eukaryotic cells which, contrary to prokaryotes, possess a nucleus separated from the remainder of the cell, the genome is the total DNA content which is enclosed within the nucleus, but some extra DNA is also found in mitochondria or blastocytes, which are organelles of the cell outside the nucleus. (They are believed to be former prokaryotes symbiotically integrated into the eukaryotic cell.) In a multicellular organism, the nucleus of each cell contains the same genome.

The nucleotides of the pair **A–T** are tied together by two hydrogen bonds, while three hydrogen bonds tie together those of the pair **G–C**. Hydrogen bonds are rather weak chemical bonds (much weaker than covalent bonds), so the double helix can fairly easily be separated into its two strands, each bearing a message complementary to that of the other one. DNA replication entirely relies on this property. Splitting the DNA molecule into its two strands, endowing each nucleotide of one strand with its complementary and constructing another 'backbone' as a support for the sequence of newly appended complementary nucleotides eventually results in two identical copies of the initial double-strand molecule. In the following, we shall refer to this replication mode as the 'template-replication paradigm' and question its ability to faithfully conserve the genetic message.

## 2.2 DNA DIRECTS THE CONSTRUCTION OF A PHENOTYPE

The message borne by the DNA molecule directs the construction of a phenotype. The best-known step in this process is how it directs the synthesis of the main constituents of living matter, complex molecules named *proteins*.

Some parts of the genome, referred to as 'genes,' control the synthesis of proteins. Proteins are made of a properly folded chain of amino-acids referred to as *polypeptidic chain*. An amino-acid is made of a carbon atom to which are attached a hydrogen atom, an amino group ($NH_2$), a carboxy group (COOH), and a specific side chain. The amino-acids of a polypeptidic chain are linked together by binding the carboxy group of an amino-acid and the amino group of another one, with elimination of a water molecule. Proteins are typically made of a few hundred amino-acids, each of them taken from a given set of as few as 20 amino-acids which differ from each other only as regards their side chain. The side chains of these amino-acids have various chemical compositions (some include a sulphur atom) and various properties, some of them being hydrophilic and others hydrophobic, and others bearing positive or negative electric charges.

The polypeptidic chain is determined by the sequence of nucleic bases of the gene. Each of its amino-acids is specified by a triplet of nucleic bases referred to as a *codon*. The order of the successive codons in the DNA strand determines the order of the corresponding amino-acids in the polypeptidic chain. The rule which tells which amino-acid corresponds to each codon is referred to as the '*genetic code*'. (We shall always use quotes when referring to it, in order to avoid any confusion with the genomic error-correcting code the existence of which we are led to assume.) After being properly folded, polypeptidic chains become proteins, which are both constituents of the living matter and actors of life processes, especially as enzymes, i.e., catalysts for the chemical reactions which take place inside the cell. An upstream 'promoter sequence' warns that a sequence of codons to be translated into amino-acids will be found. Then reading a single well-defined codon both starts the process and 'codes' for the amino-acid methionine, while reading later any of three stop codons stops it. The starting codon also determines the reading frame, i.e., what groups of three nucleotides should be dealt with as codons.

The genes constitute a part of the content of genomes which widely varies depending on the species, from almost its totality for bacteria down to a few percent, e.g., in humans, or even less. Much of the function of DNA outside the genes remains poorly understood but it is probably much more important than yet believed. In any case, DNA directs the construction of a phenotype. In turn, the phenotype hosts the genome in which it originates and its membranes shield it against mechanical and chemical aggressions.

## 2.3   FROM DNA TO PROTEIN, AND FROM A GENOME TO A PHENOTYPE

The path from codons in DNA to the corresponding amino-acids is not straightforward. To perform the synthesis of a protein, the DNA string of a gene is first copied into an RNA molecule referred to as *messenger RNA* (mRNA). This step is referred to as *transcription*. Then, the messenger RNA is processed so as to build the chain of amino-acids which will eventually become a protein. Each mRNA triplet which corresponds to a codon in DNA generates a short RNA molecule named *transfer RNA* (tRNA) which binds itself to the amino-acid which corresponds to this particular codon according to the genetic 'code'. The transfer RNAs are processed by an organelle of the cell named *ribosome* which joins together the successive amino-acids brought by transfer RNAs, forming the polypeptidic chain to be later folded into a protein. All this process is referred to as *translation*.

The paragraph above roughly describes the mechanism by which prokaryotic cells produce proteins (details of the actual process have been omitted). The mechanism is significantly more complicated in eukaryotes. In this case, the DNA string of a gene is first copied into a *premessenger RNA* which is later spliced: some of its parts named *introns* are removed from it, while the remaining parts, named *exons*, are joined together to constitute the messenger RNA which actually directs the synthesis of the polypeptidic chain. Introns and exons are reliably recognized by the molecular machinery which processes the genome but how they are recognized remains poorly understood.

In eukaryotes, only a small fraction of the DNA is actually used for directing the synthesis of proteins (it is estimated to about 1.3% of the human genome). Some of the remainder has regulatory functions, but the actual function of most of the genome remains unknown. This is why much of the genome has been dubbed 'junk DNA' and the DNA which actually directs the synthesis of proteins as 'coding DNA' (the word 'coding' referring here to the so-called genetic 'code'). The 'junk DNA' was believed to have no usefulness at all, being often interpreted as fossil or parasitic. I'll strongly suggest below that all the DNA of a genome contributes to its own conservation, hence that parasitic DNA cannot have but a short life. Recent researches show that, although a small fraction of DNA encurs both transcription and translation, hence participates in the synthesis of proteins, much of the DNA is transcribed into short RNA sequences the function of which is still poorly understood.

In multicellular organisms, although (almost) each cell possesses the genome in its entirety, only certain genes are expressed, i.e., used in order to direct protein syntheses, the other ones being 'repressed.' The genes which are expressed depend on the kind of the cell which, being differentiated, is now specialized. The gene expression is repressed by RNA molecules or proteins binding themselves at some particular places in the genome thus hampering certain regions in it to be transcribed or translated. Moreover, the expression or repression of genes depends on the particular history of the cell to which it belongs.

## 2.4    GENOMES ARE VERY LONG

Genomes are *very* long, and, moreover, their length widely depends on the species. The genome lengths range from thousands of base pairs for viruses up to hundreds of billions base pairs for certain animals and plants. The length of the human genome, for instance, is about 3.2 billion base pairs. The number of *a priori* possible genomes of a given length $n$ is $4^n = 10^{n \log_{10} 4} \approx 10^{0.6n}$. Even for the shortest genomes, this number is so large that it defies imagination. For comparison purpose, $4^{133}$ equals about $10^{80}$, the estimated number of atoms in the visible universe. The number of genomes actually associated with living beings is therefore only a tiny fraction of the total number of possible genomes. In the parlance of information theory, genomes are thus extremely *redundant*. We shall see in Sec. 3.5 and Ch. 6 that redundancy is a necessary property of error-correcting codes.

Not only all genomes are very long, but the range of their lengths is extremely large. Interestingly, the number of genes is much less variable from a specias to another (if we except viruses) since it hardly exceeds an order of magnitude. This may be thought of as an illustration of the intrinsic evolutive advantage of long genomes (as argued in Sec. 11.5). Would evolution be solely driven by natural selection acting on phenotypes, the number of genes rather than the genome length would be the most significant parameter.

A more comprehensive discussion of molecular genetics will be found in Ch. 4. Let us now begin questioning the template-replication paradigm. We saw that a phenotype hosts and shields its genome. Clearly, membranes can to some extent protect DNA against mechanical and chemical damages. But they are unable to shield it from the radiations to which the Earth habitat is subject, of terrestrial (natural radioactivity), solar and cosmic origin. Moreover, the DNA molecule is about 2-nanometer wide, so it is a quantum object in two of its dimensions. It suffers thermal noise and no deterministic description of it can be given. Both the very large size of genomes and the huge time intervals involved (the origin of life dates back to at least 3.5 billion years, and certain genes are shared by species, e.g., flies and humans, which have diverged in the evolution tree several hundred million years ago), make the conservation of the genetic message through the ages appear as almost miraculous. However, the genome is subject to many damages which occur at the much shorter time scale of individual lifetimes and are responsible for ageing or cancer. The genomic message cannot be both vulnerable to errors at a very short timescale and extremely well conserved at the timescale of geology, hence the template-replication paradigm cannot account for the faithful conservation of genomes, which is nevertheless a major biological fact. Computing the DNA capacity in Ch. 7 will dramatically confirm this observation. Faithfulness can only result from completely different mechanisms: information theory tells us that they must rely on error-correcting codes.

CHAPTER 3

# An Overview of Information Theory and Error-Correcting Codes

## 3.1 INTRODUCTION

At variance with molecular genetics, I am afraid that popular books dealing with information theory pale in comparison to that of Dawkins's. Indeed, it is especially difficult to give a fair account of information theory without having recourse to mathematical concepts and notations. For the following rough discussion of information theory, I assume that the reader is reasonably familiar with mathematics at the undergraduate level.

In the literature, I often find concepts alleged as belonging to information theory but which engineers do not recognize as such; then, 'information theory' is used as a label for topics more or less loosely related to it. I must state that in this lecture I'll restrict information theory to the *science of literal communication*, as Shannon expounded it in [72], to its further elaboration and its engineering applications. Other attempts to build information theories were made, some of them rather speculative, but only Shannon's theory has proved to be both rigorous and fully operational (if we except the algorithmic information theory to be briefly discussed in Sec. 5.4 below). Restricting information theory here to Shannon's legacy is not attempting to impose some kind of orthodoxy, but just recognizing that the engineers' experience selected it according to a kind of Darwinian process, as being extraordinarily fruitful for analyzing and designing communication devices, processes, and systems. The restriction to 'literal communication', which is indeed the purpose of communication engineering, entails that information theory ignores semantics: it discards the meaning of messages as irrelevant to their transportation. Although this postulate may seem to severely restrict the relevance of information theory, it turns out that it has been necessary for setting it up as the science of communication engineering and has never hampered its development.

Besides the formalized results of information theory, the engineers' collective experience has provided some principles which remain almost implicit. Making them explicit for a reader foreign to the field is perhaps the most difficult task when trying to give him/her a useful insight about communication engineering. According to Pierce [66], information theory brought to its practitioners wisdom rather than knowledge. I would rather say that it brought wisdom besides knowledge. Communicating wisdom is a difficult and may be a paradoxical task.

I first introduce a scheme named *Shannon's paradigm* in order to make precise the concept of communication. It involves three distinct entities: the source, the channel, and the destination. The source generates a message intended to the destination. This message necessarily passes through the channel, which is generally not faithful because perturbations prevent its output to be identical to its input. I'll then discuss how information theory quantitatively measures information as considered from the sole point of view of literal communication. It then becomes possible to associate information-theoretic quantities with the elements of Shannon's paradigm, namely its *entropy* with the source and its *capacity* with the channel. Going back to this paradigm, I'll then introduce the operations of source—and channel coding as means to transform the message generated by the source so as to make it more easily communicated. Two fundamental theorems show that the entropy and the capacity are the limits of what is possible as regards these coding processes, thus endowing these quantities with an operational significance.

## 3.2   SHANNON'S PARADIGM

The scheme of a communication shown in Fig. 3.1 is referred to as *Shannon's paradigm*. A *source* generates a *message* intended to a *destination*. A message is a sequence of events, each unambiguously signaling the choice of a particular symbol among a given finite set of symbols, named alphabet. Source and destination are distinct entities, hence spatially separated, but there exists between them a *channel* which, on the one hand, exhibits propagation phenomena such that an input to it results in an output response which can be observed by the destination; and, on the other hand, perturbing phenomena. Due to these perturbations, the input to the channel does not suffice to determine its output with certainty. The destination has no means to perceive the transmitted message but observing the channel output.

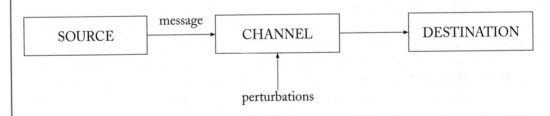

Figure 3.1:  Basic scheme of a communication: Shannon paradigm.

Notice that the scheme of Fig. 3.1 is still relevant if the source and destination are separated in time rather than in space. Propagation is then replaced by recording and the channel consists of some medium which saves long-lasting modifications specified by writing signals. Receiving consists of later reading the same medium. This interpretation of the recording and reading process as a channel in the information theoretic sense is important, since it makes information theory relevant to this case as well as to communications through space. It will enable us to deal with the DNA molecule as a channel for communicating through time.

In communication through space a feedback channel (i.e., from the destination to the source of Fig. 3.1) is often available, and many simple and efficient communication processes rely on the use of such a feedback. In communication through time, however, communicating from the future to the past is impossible since it would violate causality. No feedback channel is available, hence no such communication processes can exist. Only transmitting the message as a word of an error-correcting code and recovering it from the received message, a process referred to as 'forward error correction,' can protect the message against the channel perturbations (see Sec. 3.5 and Ch. 6).

## 3.3    QUANTITATIVE MEASUREMENT OF INFORMATION

For defining its quantitative measurement, information is considered only from the point of view of its literal communication. The basic principle of this measurement is that the more unexpected is a message, the more information it provides. Then the information quantity borne by a message is measured by some increasing function of its improbability.

### 3.3.1    SINGLE OCCURRENCE OF EVENTS

Let us first consider a single event $x$ which occurs with probability $p$. It will bring an information quantity equal to $f(1/p)$ where $f(\cdot)$ is a positive and increasing function to be properly chosen. Let us now consider two events which occur independently from each other with probabilities $p_1$ and $p_2$, respectively. Due to their assumed independence, the probability that both occur, referred to as their joint probability, equals the product $p_1 p_2$ of their individual probabilities. It is reasonable to state that their joint occurrence provides an information quantity equal to the sum of the information quantities they bear individually, so we should have

$$f(1/p_1 p_2) = f(1/p_1) + f(1/p_2).$$

This is the main reason why the logarithmic function, which satisfies this equality, is chosen for $f(\cdot)$. We denote it by $\log_\beta(\cdot)$ in order to make explicit that it depends on a positive real number $\beta$ named the 'logarithmic base,' such that $\beta^{\log_\beta(a)} = a$ for any positive real number $a$. Changing it merely multiplies the logarithm by a positive factor, hence choosing the logarithmic base defines the unit of information quantity. Shannon [72] proposed to take this base equal to 2 for measuring information, and we shall do so in the following. This choice defines the unit of information as that brought by resolving the alternative between two equally probable events. Shannon called this unit "bit" but this acronym is very often used to mean 'binary digit.' It makes generally no sense to associate an information quantity with a binary digit, most often for lack of a probability being assessed to it; even if it can be given a meaningful probability, the corresponding information quantity equals the binary information unit in the sole case where this probability equals 1/2. To avoid the use of a same word to designate the unit of information itself and an object which does not necessarily bear information and, if it does, bears the unit of information only in a particular case, the International Standards Organization (ISO) proposed the name *shannon*, abbreviated as Sh, for the binary information unit. I'll use this name for the binary information unit in the following, contrary to the current information theory literature which still most often names it 'bit.' It will be most convenient in the sequel to use the word 'bit' in order to exclusively mean 'binary symbol.'

The quantity of information brought by the single occurrence of the event $x$, with probability $p$, will thus be measured by

$$h(x) = \log_2(1/p) = -\log_2(p) \tag{3.1}$$

shannons, a positive quantity which cancels only if $p = 1$, i.e., if the considered event occurs with certainty. This is consistent with the above principle that the more unexpected is an event, the more information it provides: no information at all is brought by an event which occurs with certainty.

A single event does not suffice to describe the operation of a channel. Due to the perturbations, the output of a channel depends only probabilistically on its input, so we must consider a pair of nonindependent events: the channel input, say $x$, and the corresponding output, say $y$. We wish to measure how much information about $x$ is provided by the occurrence of the event $y$. The information quantity provided by the event $x$ is given by Eq. (3.1), but the knowledge that $y$ occurred does not suffice to determine with certainty that $x$ occurred. A residual uncertainty about $x$ remains when $y$ occurred, and it can be measured by $-\log_2[\Pr(x|y)]$, to be denoted by $h(x|y)$, where $\Pr(x|y)$ denotes the conditional probability of $x$ given that $y$ occurred. The output resolves the uncertainty about $x$, $h(x)$ as given by Eq. (3.1), except for the remaining uncertainty $h(x|y)$. The information that the occurrence of $y$ brings about $x$ may thus be measured by the *difference*:

$$i(x; y) = h(x) - h(x|y) .$$

This quantity is referred to as the *mutual information* (quantity) of $x$ and $y$. The adjective 'mutual' is relevant because this expression is actually symmetric in $x$ and $y$. Indeed, *Bayes rule* states that

$$\Pr(x, y) = \Pr(x|y)\Pr(y) \tag{3.2}$$

where $\Pr(x, y)$ denotes the joint probability of $x$ and $y$ (i.e., the probability that both occur), which is symmetric in $x$ and $y$: $\Pr(x, y) = \Pr(y, x)$). It follows that $h(x, y) = h(x|y) + h(y)$, so $i(x; y) = h(x) - h(x|y) = h(x) + h(y) - h(x, y)$, a symmetric expression. Therefore,

$$i(x; y) = h(x) - h(x|y) = h(x) + h(y) - h(x, y) = h(y) - h(y|x) = i(y; x) . \tag{3.3}$$

If $x$ completely determines $y$, then $i(x; y) = h(x)$ and no information loss results from the channel use, but if $x$ and $y$ are mutually independent, then $i(x; y) = 0$, as expected: the channel is then completely inefficient.

Actually, $i(x; y)$ as just defined is seldom used. The mean of this quantity, $I(X; Y)$, to be defined in the next section by Eq. (3.10), is on the contrary a quantity of paramount importance in information theory. It should be referred to as 'average mutual information' but for brevity's sake the word 'average' is generally dropped, so we'll use 'mutual information' in the sequel in order to name the average quantity $I(X; Y)$.

### 3.3.2   ENTROPY OF A SOURCE

Repetitive Events, Average Quantities

Events which occur only once are not truly relevant to communication engineering, and the interesting case is that of a source which repeatedly (for instance, but not necessarily, periodically) generates outputs. From now on, a source output is assumed to result from the choice of a particular element among some set of $q$ symbols named alphabet. The finite number $q$ will be referred to as the alphabet size. Instead of the information brought by a single outcome, one considers the average information quantity which is generated by the source operation. It has been named by Shannon the *entropy* of the source, but its relation with the entropy of physics is not straightforward and will not be discussed here. We moreover assume to begin with that the successive outputs are mutually independent. Then the source is referred to as 'memoryless.'

Entropy of a Memoryless Source

Let us first define a finite *random variable* for describing the output of a memoryless source. It is a quantity, say $X$, which can assume a finite number of different values, or outcomes, say $x_1, x_2, \ldots, x_q$, each occurring with some definite probability: $\Pr(X = i) = p_i$, for $i = 1, 2, \ldots, q$. The probabilities associated with all outcomes sum up to 1 since one of the outcomes occurs with certainty:

$$\sum_{i=1}^{q} p_i = 1 \,. \tag{3.4}$$

It is usual and convenient to represent a random variable by a capital letter and a particular value it assumes by a lowercase one. For instance, $X = x$ or $X = 1$ means that the random variable $X$ is given the particular value $x$ or 1. If the source output is the random variable $X$, then its entropy $H(X)$ is the average of $h(\cdot)$ as given by Eq. (3.1), namely:

$$H(X) = \sum_{i=1}^{q} p_i \log_2(1/p_i) = -\sum_{i=1}^{q} p_i \log_2(p_i) \,. \tag{3.5}$$

Here, and in the following, we denote average information quantities by capital letters. As a particular case, the entropy of a binary random variable where the symbol probabilities are $\Pr(0) = p$ and $\Pr(1) = 1 - p$ is

$$\mathcal{H}_2(p) = -p \log_2(p) - (1 - p) \log_2(1 - p) \,, \tag{3.6}$$

which is represented in Fig. 3.2 as a function of $p$.

Let us give some examples of sources and calculate their entropy. A binary source with equiprobable symbols, say **S1**, i.e., such that $\Pr(0) = \Pr(1) = 1/2$ has an entropy of 1 shannon: $H_{S1} = \mathcal{H}_2(1/2) = -(1/2) \log_2(1/2) - (1/2) \log_2(1/2) = 1$. Unequal symbol probabilities decrease the entropy. For instance, source **S2** for which $\Pr(0) = 0.9$ and $\Pr(1) = 0.1$ has as entropy $H_{S2} = \mathcal{H}_2(0.9) = -0.9 \log_2(0.9) - 0.1 \log_2(0.1)$ which approximately equals 0.469 Sh. A quaternary source with equiprobable symbols ($\Pr(0) = \Pr(1) = \Pr(2) = \Pr(3) = 1/4$), say **S3**, has an entropy of 2 Sh.

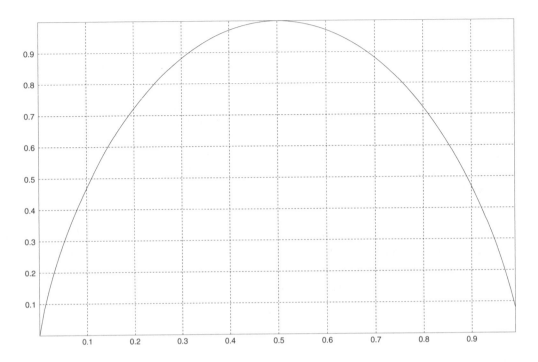

**Figure 3.2:** Binary entropy function $\mathcal{H}_2(p)$.

Entropy of a Source with Memory

It is necessary to extend the concept of entropy to sources with memory, i.e., where the successive output symbols are mutually dependent. The entropy of such a source can be defined as the limit for $s$ approaching infinity of $H_{\underline{c}}(s)/s$, where $\underline{c}$ denotes a sequence of source output symbols of length $s$. $H_{\underline{c}}(s)$ is defined as:

$$H_{\underline{c}}(s) = -\sum_{\underline{c}} \Pr(\underline{c}) \log_2[\Pr(\underline{c})] \,, \tag{3.7}$$

where $\Pr(\underline{c})$ denotes the probability that sequence $\underline{c}$ occurs, and the sum in the above expression is extended to all possible output sequences of length $s$. The entropy of a source with memory is thus defined as:

$$H(X) = \lim_{s \to \infty} \frac{1}{s} H_{\underline{c}}(s) \,. \tag{3.8}$$

This limit exists provided the source is stationary, i.e., provided the probabilities which describe its operation remain constant. An especially important application of this extension will be found in the case of error-correcting codes.

The simplest examples of sources with memory are Markovian. For instance, a binary Markovian source has been represented in Fig. 3.3.

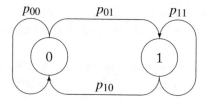

**Figure 3.3:** A binary Markovian source.

When the source output is $x$, the probability that the next symbol is $y$ is $\Pr(y|x)$, denoted in the figure and below by $p_{xy}$, where both $x$ and $y$ are binary symbols 0 or 1; $p_{01}$, for instance, is the probability that the output 1 follows the output 0. Obviously, $p_{xy} + p_{x\bar{y}} = 1$, where $\bar{y}$ denotes the binary complement to $y$. The source is thus fully defined by two independent probabilities, e.g., $p_{01}$ and $p_{10}$. Although the source operation depends to some extent on the initially transmitted symbol, its influence vanishes as time passes and a steady state is reached where the symbols 0 and 1 occur with constant probabilities referred to as steady or stationary, which are easily shown to be

$$\Pr(0) = \frac{p_{10}}{p_{10} + p_{01}}, \quad \Pr(1) = 1 - \Pr(0) = \frac{p_{01}}{p_{10} + p_{01}}. \tag{3.9}$$

Notice that if $p_{10} = p_{01}$, we have $\Pr(0) = \Pr(1) = 1/2$ as for source **S1**, although the probability distribution of the generated sequences generally differs from that of **S1**.

The entropy of the above Markovian source is the average entropy of the two states 0 and 1, i.e.,

$$H = -\Pr(0)[p_{00} \log_2(p_{00}) + p_{01} \log_2(p_{01})] - \Pr(1)[p_{10} \log_2(p_{10}) + p_{11} \log_2(p_{11})],$$

or, according to Eq. (3.9):

$$H = -\frac{p_{10}}{p_{10} + p_{01}}[p_{00} \log_2(p_{00}) + p_{01} \log_2(p_{01})] - \frac{p_{01}}{p_{10} + p_{01}}[p_{10} \log_2(p_{10}) + p_{11} \log_2(p_{11})].$$

In the following, we shall consider as examples the binary Markovian sources where $p_{10} = p_{01} = 0.1$ (say **S4**), and where $p_{10} = 0.9$ and $p_{01} = 0.2$ (**S5**). The entropy of source **S4** is about 0.469 Sh, and that of **S5** is about 0.676 Sh.

The Information Measures Depend on the Ascribed Probabilities

In Sec. 3.3.1 above, we assumed that a probability could be ascribed to a single event, while in Sec. 3.3.2 a probability has been ascribed to each of the outcomes of the source, the output of which is thus a random variable with this probability distribution. Saying that the information measures defined above depend on the ascribed probabilities is merely stating the obvious, but it needs to be emphasized since its consequences are very often forgotten. It should first be recalled that it is not always possible to ascribe a probability to an event. Second, arbitrarily ascribing a probability to an event results in an information measure which is no less arbitrary than this probability itself. In any case, the information measures are relative to the probabilities ascribed to events and by no means absolute.

The probabilistic information measures apply well to situations where highly repetitive events occur with stable frequencies. Then the probabilities can rightfully be estimated from frequency measurements. It is why the entropy of a source as defined in Sec. 3.3.2 can be considered as experimentally accessible if certain conditions are met (stationarity and ergodicity; see Secs. 5.1.1 and 5.1.2). Then meaningful probabilities can be ascribed to the source outputs. It should be emphasized, however, that the information borne by a single event has no meaning unless the source from which it originates is specified, i.e., it solely depends on the set of events to which it belongs and on the probabilities which are ascribed to them. We now illustrate this important point by examples concerning outputs from sources **S1** to **S5** above.

Consider, for instance the occurrence of the symbol 0. If it has been generated by source **S1** or by source **S4**, its probability is 1/2 so it bears an information of 1 Sh. For source **S2**, however, its probability is 0.9, so the corresponding quantity of information is $-\log_2(0.9)$, i.e., approximately 0.152 Sh. From source **S3**, the probability of occurrence of the same symbol 0 is 1/4 so it now bears 2 Sh. The occurrence of 0 from source **S5** occurs with the steady probability of 0, equal to 9/11 according to Eq. (3.9), and thus corresponds to approximately 0.290 Sh.

Now consider the occurrence of the 4-symbol sequence 0101. As an output of source **S1**, this event has as probability 1/16 and the corresponding information quantity is 4 Sh. From source **S2**, however, its probability equals $0.9 \times 0.1 \times 0.9 \times 0.1 = 0.0081$ and the corresponding information quantity is approximately 6.948 Sh. If it is generated by the quaternary equiprobable source **S3**, the same sequence occurs with probability 1/256 hence bears an information quantity of 8 Sh. For the Markovian source **S4**, the probability of generating 0101 is $0.5 \times 0.1 \times 0.1 \times 0.1 = 1/2000$ because the probability of the first 0 is the steady probability of zero, while the conditional probability that it is followed by 1 is 0.1, that this 1 is followed by 0, and then that 0 is followed by 1 are again 0.1. The corresponding information quantity is thus approximately 10.966 Sh. For source **S5**, the probability of this sequence is, similarly, $9/11 \times 0.2 \times 0.9 \times 0.2$ and the corresponding information quantity is approximately 5.085 Sh.

The examples above clearly show that the information borne by a message does not depend on this message itself, but on the set of messages from which this particular message is taken. A source generates such a set of messages, and its entropy, i.e., the information-theoretic significant quantity associated with it, only depends on the probabilities ascribed to the elements of this set. The measure of information is only meaningful for events endowed with known probabilities, a stringent condition which limits the usefulness of Shannon's information theory. Failing to clearly identify the relevant probabilistic set of events often nullifies statements allegedly using information-theoretic measures.

The *algorithmic information theory* is a possible alternative to Shannon's information theory, originated in computer science, to be briefly discussed in Sec. 5.4. It does not base the measure of information on probabilities, which makes it of great conceptual interest. Its practical usefulness in communication problems is however limited, and it is why this lecture relies on Shannon's probabilistic measure of information. It should moreover be realized that the algorithmic information theory results in essentially the same information measures as Shannon's in instances where both theories are relevant, at least for long enough sequences. Rather than competing, the algorithmic information theory and Shannon's appear as two facets of a same science of information.

### 3.3.3    AVERAGE MUTUAL INFORMATION, CAPACITY OF A CHANNEL
Average Mutual Information

Let us now consider the average of the mutual information $i(x; y)$ between the input $x$ and the output $y$ of a channel, given by Eq. (3.3). We obtain:

$$I(X; Y)=H(X)-H(X|Y)=H(X)+H(Y)-H(X, Y)=H(Y)-H(Y|X)=I(Y; X), \quad (3.10)$$

where the entropy $H(X)$ has already been defined in Eq. (3.5). The joint entropy $H(X, Y)$ is the straightforward extension of the entropy definition to a pair of variables, namely:

$$H(X, Y) = - \sum_{i=1}^{q} \sum_{j=1}^{r} \Pr(x_i, y_j) \log_2[\Pr(x_i, y_j)] ,$$

assuming that $Y$ has as possible outcomes $y_1, y_2, \ldots, y_r$, and the conditional entropy $H(X|Y)$ is similarly the average of $h(x|y)$, namely:

$$H(X|Y) = - \sum_{i=1}^{q} \sum_{j=1}^{r} \Pr(x_i, y_j) \log_2[\Pr(x_i|y_j)] .$$

Notice that the coefficients in the above expression are the *joint* probabilities of $X$ and $Y$, as associated with a pair of random variables, while the probabilities in argument of the logarithms are *conditional* probabilities. Conditioning decreases the prior uncertainty hence reduces the entropy, so the inequality $H(X|Y) \leq H(X)$ implies that $I(X; Y)$ is positive (zero if and only if $X$ and $Y$ are independent).

The above expressions of $H(X, Y)$ and $H(X|Y)$, as well as the expression Eq. (3.10) of the average mutual information, are valid for a memoryless source and a memoryless channel. They can, in principle, be extended to sources and channels with memory according to Eq. (3.8) and Eq. (3.7). The corresponding expressions can be very complicated, but the important fact is that the relevant quantities are perfectly defined, hence exist in the mathematical meaning of this word under comparatively mild conditions.

Capacity of a Channel

The average mutual information depends on both the source and the channel. In order to obtain a quantity which solely characterizes the channel, one defines its *capacity* as the average mutual information obtained by choosing the source in order to maximize it. It is thus the largest possible average mutual information that the channel can convey. The channel capacity cannot of course exceed the maximum entropy of a source connected to its input, hence cannot exceed $\log_2 q$ shannons when $q$ is the size of the channel input alphabet. The channel perturbations actually diminish the capacity.

Let us give two examples of channels and of their capacity. For simplicity's sake, we restrict ourselves to memoryless channels. In the following figures, a memoryless channel is represented by the set of arrows associated with the transitions from the symbols of the input alphabet to symbols of the output alphabet. Each arrow is labeled with the probability that the transition it represents occurs. The channels are memoryless in the sense that each transition occurs independently from the previous ones. The figure at left depicts the *binary symmetric channel*, that at right the *binary erasure channel*.

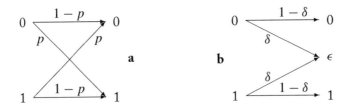

**Figure 3.4:** Binary-input channels. On the left, binary symmetric channel with error probability $p$. On the right, binary erasure channel , with erasure probability $\delta$.

The binary symmetric channel is depicted in Fig. 3.4-**a**. Both its input and output alphabets are binary. The word 'symmetric' means here that the probability $p$ of a 0 being transformed into a 1 equals that of a 1 being transformed into 0. We shall refer later to this kind of error as 'substitution.' It turns out that the mutual information of a symmetric channel is maximized when the source is memoryless and equiprobable, i.e., the probability of occurrence of 0 equals that of 1, so both probabilities equal 1/2. We have also, by symmetry, $\Pr(Y = 0) = \Pr(Y = 1) = 1/2$. One thus has $H(X) = H(Y) = 1$ shannon. The joint probabilities are

$$\Pr(X = 0, Y = 0) = \Pr(X = 1, Y = 1) = (1 - p)/2$$

and

$$\Pr(X = 0, Y = 1) = \Pr(X = 1, Y = 0) = p/2 \,,$$

hence

$$H(Y|X) = -(1 - p) \log_2(1 - p) - p \log_2(p) = \mathcal{H}_2(p) \,,$$

where the function $\mathcal{H}_2(p)$ is the entropy of a binary random variable which assumes one of its values with probability $p$, as already defined in Eq. (3.6). The capacity of the binary symmetric channel, i.e., $I(X; Y)$ for $\Pr(X = 0) = \Pr(X = 1) = 1/2$, is thus:

$$C_{\text{bsc}} = H(Y) - H(Y|X) = 1 - \mathcal{H}_2(p) \,. \tag{3.11}$$

One notices that $C_{bsc} = 0$ for $p = 1/2$. In this case, the channel output has become independent of its input, making the channel useless. The capacity achieves its maximum value of 1 Sh for $p = 0$ and $p = 1$. If the first case is not surprising, the second one is a bit more, but it suffices to swap the labels '0' and '1' of the output symbols (which are in fact arbitrary) to revert to the case where $p = 0$. Due to symmetry, no restriction of generality results from assuming that $p \leq 1/2$.

For the binary erasure channel, which is represented in Fig. 3.4-**b**, the input alphabet is binary but the output alphabet is ternary, its symbols being denoted by 0, 1, and $\epsilon$. Transitions can occur from an input 0 to an output 0 or $\epsilon$, and from an input 1 to an output 1 or $\epsilon$, but never from 0 to 1 or vice versa. The probability of a transition from 0 to $\epsilon$ is the same as that from 1 to $\epsilon$ and is denoted by $\delta$. The third output symbol $\epsilon$ may be interpreted as indicating that the output symbol has not been recognized as belonging to the input alphabet, hence has been 'erased.' This channel is symmetric, too, so calculating its capacity still consists of computing $I(X; Y)$ for $\Pr(X = 0) = \Pr(X = 1) = 1/2$. The joint probabilities are then:

$$\Pr(X = 0, Y = 0) = \Pr(X = 1, Y = 1) = (1 - \delta)/2$$

and

$$\Pr(X = 0, Y = \epsilon) = \Pr(X = 1, Y = \epsilon) = \delta/2 ,$$

all others being zero. It then results that

$$\Pr(Y = 0) = \Pr(Y = 1) = (1 - \delta)/2$$

and

$$\Pr(Y = \epsilon) = \delta ,$$

hence

$$H(Y) = -(1 - \delta) \log_2[(1 - \delta)/2] - \delta \log_2(\delta) = 1 - \delta + \mathcal{H}_2(\delta) ,$$

and

$$H(Y|X) = -(1 - \delta) \log_2(1 - \delta) - f \log_2(\delta) = \mathcal{H}_2(\delta) .$$

The capacity $C_{ec}$ has thus the very simple expression:

$$C_{ec} = H(Y) - H(Y|X) = 1 - \delta . \tag{3.12}$$

It may be interpreted as follows: in the average, a fraction $\delta$ of the symbols is lost. Each of the remaining ones bears an information quantity of one shannon, so the average information quantity borne by the channel output is $1 - \delta$ Sh per input binary symbol.

## 3.4    CODING PROCESSES

### 3.4.1    VARIANTS OF SHANNON'S PARADIGM

Let us go back to Shannon's paradigm as depicted in Fig. 3.1. We now introduce variants by including in this scheme devices intended to transform the message.

If the source, the channel, and the destination are arbitrary, nothing *a priori* guarantees the mutual compatibility of the source and the channel, on the one hand, and of the channel and the destination, on the other hand. For instance, in radiotelephony, the source and destination are human beings, but the channel is provided by the ability of electromagnetic waves to propagate. Humans, however, have no natural means for transmitting and receiving such waves, except that we perceive light radiations. We thus have to augment the scheme of Fig. 3.1 with devices intended to match the blocks of the initial scheme to each other. One then obtains the scheme of Fig. 3.5-a. It is a mere variant of the scheme of Fig. 3.1, since the set comprising the source and the transmitting device on the one hand, that comprising the receiving device and the destination on the other hand, may be interpreted as a new source-destination pair now matched to the initially given channel (Fig. 3.5-b). One may as well consider the set comprising the transmitting device, the channel, and the receiving device as being a new channel, matched to the original source-destination pair (Fig. 3.5-c); thus, in the previous example we considered as a channel a radiotelephonic link, made of the set of transmitting devices, the propagation medium, and the set of receiving devices.

A more fruitful point of view actually consists of splitting each of the transmission and reception apparatuses into two blocks. One of the blocks of the transmitting device is matched to the source and the other one to the channel input. Similarly, the receiving device is split into a block matched to the channel output and another one matched to the destination. Interestingly, this scheme enables *normalizing* the characteristics of the blocks of Fig. 3.5, redefined as follows: new source before point A in Fig. 3.5-d; new channel between points A and B; new destination beyond point B. The engineering problems then consist of *separately* designing the pairs of matching blocks denoted in the figure by TA1 and RA1 on the one hand, by TA2 and RA2 on the other hand. What this normalization consists of will be precisely stated later.

In general, the borders of the blocks in Fig. 3.5 may be freely redefined for the sake of analysis; cutting any apparatus chain linking a source and a destination in two points such that the origin of the message useful to the destination is in the leftmost block defines a new triplet source-channel-destination, *provided* the *perturbations* only affect devices located in the *central* block, i.e., the channel. To prevent misunderstandings, let us stress that the concepts of information and perturbation are by no means absolute. On the contrary, the message from the source is considered as bearing information only because it can be used by the *destination*. The perturbing events are just those which are harmful with respect to this particular choice of the information message. The usefulness to the destination is the *only* criterion which differentiates the information message from the perturbations, meaning that their distinction is *arbitrary*. They may well exchange their roles. For example, the sun is a source of noise for a receiver intended to the signal of a communication satellite. However, the satellite signal perturbs the observation of a radioastronomer who studies solar electromagnetic radiations, hence it generates perturbations for this particular destination. When we refer to the usefulness or harm that a sequence of symbols has for the destination, we necessarily consider its *purpose*, hence we cannot avoid some kind of teleonomy or subjectivity.

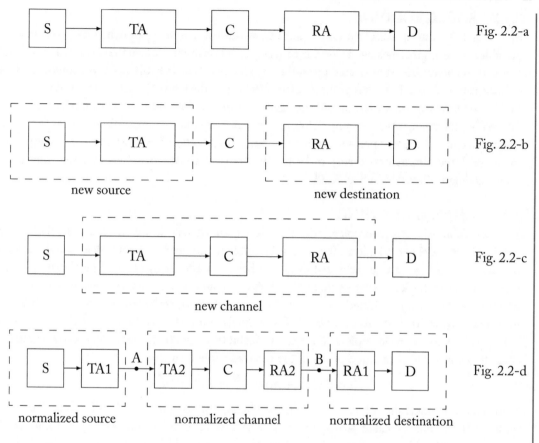

**Figure 3.5:** Variants of Shannon's paradigm. S stands for "source," C for "channel," and D for "destination." TA means "transmission apparatus" and RA, "reception apparatus."

It is possible to transform any message into a strictly equivalent one which possesses some properties more favourable to its communication than the original ones. Such transformations are referred to as coding processes. I'll briefly discuss the main two ones, namely *source coding* and *channel coding*. The information-theoretic quantities which have been associated with the blocks of Shannon's paradigm, namely the *entropy* of the source and the *capacity* of the channel, then appear as the limits of what is possible when the encoding operations are ideally performed. These two quantities acquire therefore an operational significance, which confirms their adequacy to communication problems and enlightens their meaning.

### 3.4.2 SOURCE CODING

Source coding aims at *maximum conciseness*, i.e., at substituting an equivalent message as short as possible for the original message according to a reversible transformation, in the sense that the original message should exactly be recovered given the encoded one. Source coding is a very important topic of information theory, but it has probably little relevance to the field of genetics. It seems indeed that DNA is quite 'cheap' since even the shortest genomes are very long, so there is no obvious biological interest in performing source coding. Although it is unlikely that nature implements source coding, it is a concept extremely important for understanding information theory as a whole, and especially the algorithmic information theory to be briefly discussed in Sec. 5.4. It is why the principle of source coding is discussed in Sec. 5.2.3.

### 3.4.3 CHANNEL CODING

The aim of *channel coding* is completely different: it is intended to faithfully communicate the message despite the channel perturbations. It replaces the original message by an equivalent one from which the original message can be recovered even if it has been altered by channel perturbations. As a very simple example, let us assume that we wish to communicate a binary message within a channel where errors occur: a transmitted 0 is received as 1, or vice versa, with some nonzero probability $p_e$, for instance $p_e = 10^{-3}$. We transmit each of its symbols in triplicate and take a decision by majority voting among the received triplets. Then the probability $p_d$ of an erroneous 'decoding decision' is equal to the probability of 2 or 3 errors in the received triplet, namely:

$$p_d = 3p_e^2(1 - p_e) + p_e^3 = 3p_e^2 - 2p_e^3 .$$

With an error probability of $p_e = 10^{-3}$ in the channel, we obtain that the decoding decision is wrong with the much smaller probability $p_d = 2.998 \; 10^{-6}$. A still smaller error probability would have been obtained by repeating 5, 7, ... times the transmitted binary symbols, the decision rule remaining majority voting.

This example shows both that it is possible to protect a message against perturbations by coding, and the price to pay for this: an increase in the message length referred to as *redundancy*. This is only a rudimentary coding process and similar results could have been obtained at the expense of much less redundancy, by using more sophisticated 'error-correcting codes'. It is nevertheless plainly true that protecting a message against the channel perturbations necessarily implies introducing redundancy.

The aims of source coding and channel coding are thus incompatible. There exists even an antagonism (or complementarity) between them, since source coding actually increases the vulnerability to errors as well as it increases conciseness. In source coding, a single binary digit of the coded message being in error may result in a large fraction of the recovered message to be wrong. This simple remark shows that decreasing redundancy and protecting against errors may not be considered independently of each other.

As regards genetic applications, and contrary to source coding, channel coding is crucial as being the only possible way for ensuring the faithful conservation of genomes.

### 3.4.4    NORMALIZING THE BLOCKS OF SHANNON'S PARADIGM

It now becomes possible for us to make precise what should be intended by the normalization of the blocks of Fig. 3.1. The message generated by the source is first transformed by *source coding* into a message (ideally) *devoid of redundancy*, i.e., achieving the largest possible entropy, where all successive symbols are independent and where all the alphabet symbols occur with the same probability. The encoding thus performed only matches the source characteristics. Its result is highly vulnerable to perturbations since any of its symbols is essential to the message integrity. *Channel coding* then becomes mandatory so as to make the message generated by the source encoder (ideally) invulnerable to the channel perturbations, which necessarily demands reintroducing redundancy.

We may assume that source coding has been performed ideally. Then, channel coding just protects a redundancy-free message against the channel perturbations. If the message to be encoded is not completely devoid of redundancy, the protection achieved is at least that for a redundancy-free message: exploiting some remaining redundancy can be but beneficial to the protection against channel perturbations.

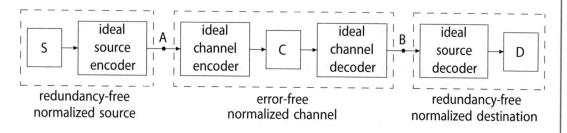

**Figure 3.6:** Normalization of the blocks 'source,' 'channel,' and 'destination' of Shannon's paradigm. S, C, and D denote the original source, channel, and destination, respectively.

One may then redraw Fig. 3.5-d as in Fig. 3.6, where the normalized source generates a redundancy-free message and where the normalized channel is error free.

The approach which consists of suppressing the redundancy of the original message by source coding, and then of reintroducing redundancy by channel coding, may look self-contradictory but the kind of redundancy of the original source is not necessarily well adapted to the properties of the channel to which it is connected. Rather than globally designing coding systems in order to match a particular source to a particular channel, the normalization as just defined enables dealing with the source and the channel separately, instead of the source-channel pair. This normalization has also an auxiliary advantage: the message alphabet at points A and B of Fig. 3.6 is arbitrary. We may thus assume it to be the simplest possible, i.e., binary, without significantly restricting generality.

### 3.4.5    FUNDAMENTAL THEOREMS

To summarize, coding operations ideally result in:

— a redundancy-free (or maximum-entropy) message, although the original message is generated by a redundant source, as regards source coding;

— an error-free message after decoding, although the coded message is received through a perturbed channel, as regards channel coding.

These possibilities are stated by the *fundamental theorems* of information theory, under conditions they precisely define. Their proof does not demand that the means for obtaining these results are made explicit. In fact, except for some simple cases, the optimum coding means remain unknown especially as regards channel coding.

The fundamental theorem of *source coding* tells that it is possible to eliminate all redundancy from a stationary source. Coding should be effected on the $k$-th extension of this source, for $k$ large enough, the stated result being possibly reached only asymptotically. The $k$-th *extension* of a source $S$ with an alphabet of size $q$ (this source being said $q$-ary) is the $q^k$-ary source which results from the original source by grouping its symbols by blocks of $k$, each of these blocks being interpreted as a symbol of the $q^k$-ary alphabet. Denoted by $S^k$, this extension is just another way of representing $S$, not a different source. When optimized, the resulting message has an *average length* per symbol of the original source which tends to its *entropy* expressed using $q$ for defining the information unit, i.e., using logarithms to the base $q$. Efficient algorithms which perform source coding are known for long and, for simple enough sources, their performance is close to the theoretical limit. Moreover, they can easily be made *adaptive* so as to deal with sources of slowly varying or even *a priori* unknown parameters.

The fundamental theorem of *channel coding* is asymptotic as the length $n$ of the words associated with blocks of $k$ source symbols tends to infinity. Keeping constant the code rate $R = k/n$ demands that $k$ increases proportionally to $n$, i.e., that larger and larger blocks of the message that the source generates are encoded. (This statement is not the most general possible, but only concerns the class of so-called 'block codes', and it is moreover assumed here that the channel input and output alphabets are the same.) The necessary condition for the error probability to approach 0 as $n$ increases can be stated as follows: the source *entropy* should be *less* than the channel *capacity* (source and channel are assumed to satisfy proper regularity conditions which are mandatory to guarantee the existence of these quantities). Thus, the presence of channel perturbations does not limit the reliability with which the message is communicated, as measured for instance by how small is the probability that it is not correctly recovered (to be referred to in the sequel as residual error probability), but only the rate at which information can be communicated using this channel. The highly desirable result of an arbitrarily small error probability can however be obtained only when using an appropriate code, for which no practically usable construction process is proposed by information theory. At least from a practical point of view, the turbocodes (invented in 1993) exhibit small residual error probability at rates close to the channel capacity so they can be considered as approaching the theoretical limit as closely as desired. They also provide an experimental evidence that the capacity is a very sharp limit and strikingly confirm the relevance of information theory, and especially its concept of channel capacity, to the communication problems.

# 3.5    A BRIEF INTRODUCTION TO ERROR-CORRECTING CODES

## 3.5.1    REDUNDANT CODE, HAMMING DISTANCE, AND HAMMING SPACE

Let us consider the set $S_n$ of $n$-symbol binary sequences. Their total number is $2^n$. A redundant code $\mathcal{C}(n, k)$, with $k < n$, is a subset of $S_n$ which contains $M = 2^k$ elements, to be referred to as codewords. We may define a one-to-one correspondence between each of these codewords and each of the possible $2^k$ binary messages. Transmitting one of the codewords is thus equivalent to transmitting one of these messages, and we assume that only codewords of $\mathcal{C}(n, k)$ are actually transmitted. Since $k < n$, the proportion of the words in $\mathcal{C}(n, k)$ among the elements of $S_n$ is $2^k/2^n = 1/2^{n-k}$: they are only a minority subset of $S_n$ if $n - k$ is large enough. For instance, we may assume as an order of magnitude that $k = n/2$. In this case, the proportion of sequences of $S_n$ which belong to $\mathcal{C}(n, k)$ is $1/1,024$ for $n = 20$, about $8 \times 10^{-31}$ for n=200 and $9 \times 10^{-302}$ for $n = 2,000$. For large values of $n$ and if the code rate $R = k/n$ is kept smaller than 1, only a tiny fraction of the $n$-symbol binary sequences belongs to the code.

I just defined a redundant code as a subset of the set $S_n$ of the $n$-symbol binary sequences. The word 'code' can be used in at least two other different meanings. As a subset of $S_n$, $\mathcal{C}(n, k)$ is generally defined by constraints specifically satisfied by its words, and I'll occasionally use the word 'code' for referring to this set of constraints. The encoding rule which establishes a correspondence between the $2^k$ information messages and the codewords is also sometime referred to as a 'code' but I'll avoid doing so.

The restriction to binary sequences is unnecessary, and an alphabet of any size can be used. Also notice that we defined redundancy by the scarcity of the codewords among all the sequences of $S_n$. This definition of redundancy is much more comprehensive than the usual reference to repeated parts. It simply means that the number of symbols $n$ in a codeword is larger than the number $k$ of message symbols which strictly suffices to specify it. The restriction to memoryless sources and channels was just for simplicity's sake, and the above concepts and results generalize to source and channels with finite memory, at the expense of an increased complexity. We may think of each of the $k$ information symbols as chosen at random with uniform probability $(1/q)$ among the $q$ alphabet symbols independently of the others, a choice which maximizes the information quantity they bear. Then, the total information (expressed in shannons) borne by an $n$-symbol word is $k \log_2 q$ where $q$ is the alphabet size, so the entropy per symbol of the encoded word is $H = k \log_2 q/n = R \log_2 q$. The condition of the fundamental theorem that the source entropy should be less than the channel capacity becomes $R \log_2 q < C$.

When the encoding rule is such that the $k$ information symbols appear explicitly at $k$ defined positions in the encoded word (e.g., but not necessarily, in the first $k$ positions), the code is said to be in *systematic form*. This is a feature of the encoding rule and not of the code as a set of words, so the often-used term 'systematic code' to designate this case is an abuse of language.

### 3.5.2   RECEPTION IN THE PRESENCE OF ERRORS

Assume again for simplicity's sake that the code alphabet is binary. Suppose now that the codeword transmitted in order to represent some message suffers errors in the sense that the symbols are transformed into their complements, i.e., 0 into 1 or 1 into 0, with some nonzero probability. If $n - k$ is large enough, the probability that a codeword is transformed into another one is very small due to the extreme scarcity of the codewords within the set $S_n$. Checking whether a received $n$-symbol binary sequence belongs or not to the code is thus a very efficient means of checking whether one or more symbol errors affected the transmitted sequence which, by hypothesis, belongs to the code $\mathcal{C}(n, k)$. Then the code enables *error detection*. Let us now introduce a metric into $S_n$, i.e., define a distance between its elements. The *Hamming distance* between two sequences is defined as the number of positions in the sequence where their symbols differ (for instance, the Hamming distance between the two 7-symbol words 1101000 and 0110100 is 4). The set $S_n$ endowed with this metric will be referred to as the $n$-dimensional Hamming space. Two different sequences of length $n$ are at a distance of at least 1 and at most $n$. Since the codewords are very sparse among the sequences of $S_n$ for $n - k$ large enough, we may choose them so as to make the smallest distance between any codeword and all the other ones as large as possible. This smallest distance $d$ is referred to as the *minimum Hamming distance* of the code. If $e$ symbols are affected by errors, the received sequence is at a distance $e$ from the transmitted word, while it is at a distance of at least $d - e$ from any other codeword. The set of locations of the symbol errors is referred to an *error pattern* and the number $e$ of these errors is referred to its *weight*. Provided $e < d - e$, or $e < d/2$, the codeword actually transmitted is thus the closest to the received sequence so it can be unambiguously determined despite the occurrence of $e$ errors in the received $n$-symbol sequence.

The process of decoding can thus be decomposed into two steps: the first and most complex one, the 'codeword recovery,' consists of looking for the word of the code the closest to the received sequence. We may name it *regeneration* since it is intended to restore the original encoded sequence. In the presence of random errors it can fail with some nonzero probability hence result in a codeword different from the transmitted one. The second step, the 'message recovery,' delivers the message associated with the codeword obtained in the first step. The encoding rule being a one-to-one correspondence, this is a trivial task which can be performed by merely reading an entry in a table. Figure 3.7 represents decoding as a two-step process.

Provided the minimum distance of the code is large enough with respect to the codeword length, the probability of an erroneous decoding decision can be made as low as desired by using appropriate codes of increasing length and adequate decoding algorithms. Then, within these limits, error-correcting codes perform the paradoxical function of enabling *reliable communication through unreliable channels*. 'Reliable' means that it is possible to make the residual error probability as low as desired, thus freeing the communication faithfulness from the perturbing events which occur in the channel. In other words, contrary to intuition, the existence of errors in the channel does not hinder communication to be performed with an arbitrarily small error probability provided an efficient and redundant enough error-correcting code is used.

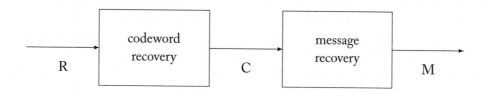

**Figure 3.7:** Decoding as a two-step process. R stands for 'received sequence,' C for 'codeword,' and M for 'message.'

Notice that any symbol in a codeword contributes to the correction of errors, hence to its own conservation. No parts of genomes, acting as codewords according to our main hypothesis, can be superfluous and the idea of 'junk DNA' is a fallacy.

## 3.6    A VARIANT OF SHANNON'S PARADIGM INTENDED TO GENETIC APPLICATIONS

In Figs. 3.1 and 3.6, the source is clearly identified as the origin of the message. Sending an information message is intentional when the source is a human being. We now propose a variant of Shannon's paradigm which better fits the absence of intentionality which is typical of the biological context (where, at least, we do not know if there is an intention).

In accordance with a remark above, we first remove from Fig. 3.6 the source encoder and source decoder which are unlikely to exist in genetics. We thus only keep the channel encoder and decoder, simply referred to as 'encoder' and 'decoder.' Then we merge the two blocks 'source' and 'encoder' into a single block (inside a dashed box) labeled 'redundant source' and, similarly, the two blocks 'decoder' and 'destination' into a single one (again inside a dashed box) labeled 'redefined destination.' We obtain the variant of Shannon's paradigm depicted in Fig. 3.8.

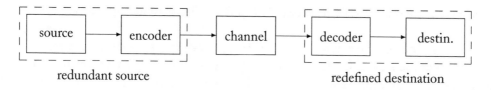

**Figure 3.8:** A variant of Shannon's paradigm.

The inclusion of the channel encoder in the redundant source endows it with error-correcting capabilities which can be exploited by the redefined destination. The entropy of the redundant source, according to the extension of the entropy concept to sources with memory, is $H = \log_2(q)k/n$ shannons, where $q$ is the alphabet size, $k$ the number of information symbols of the message, and $n$

the code length. According to the fundamental theorem of channel coding, errorless communication is theoretically possible provided the entropy $H$ is less than the channel capacity $C$.

Let us now consider the case of a genetic message incurring successive regenerations. The scheme at top of Fig. 3.9 depicts the case where two successive channels are used. Cascading the decoder which follows the first channel and the encoder which precedes the second channel results, according to Fig. 3.7, in the box labeled 'message revovery' of the decoder being followed by the encoder. But the latter just undoes what the former operated, so they cancel each other: both can be omitted. We may replace the decoder followed by the encoder by a single box labeled 'regenerator' where the encoded message no longer explicitly appears, as depicted at bottom of Fig. 3.9.

The concept of *regeneration* is better fitted to the context of genetics than the engineering concept of decoding which refers to an explicit 'information message' to be communicated. Regeneration is a necessary step, the most important indeed, of the error-correction process: identifying the presumed transmitted codeword is a nontrivial task which incurs a risk of failure.

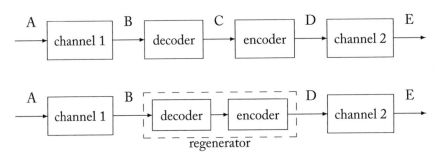

**Figure 3.9:** The regeneration function. Channels 1 and 2 have to be used successively. The upper picture is relevant to a conventional error-correcting code. The sequences found at the points designated by letters are: in A, an encoded sequence; in B and E, received sequences; in C, an information message; and in D, the sequence which results from the encoding of the decoded information message, hence restoring the initial sequence if no decoding error occurs. In the lower picture, the decoder and encoder have been merged into a single entity labeled 'regenerator' where the information message no longer appears.

We may now describe a chain of successive regenerations performed on a recorded message (e.g., the genetic message of DNA) as depicted in Fig. 3.10. An original redundant source at left delivers a redundantly encoded message which is written on a first support (labeled 'channel 1'), regenerated in the device labeled 'regen. 1,' written again on 'channel 2,' etc. The last step considered is the $i$-th regeneration, where $i$ is a finite but possibly large number.

The redundant source has been defined in Fig. 3.8 and regenerators in Fig. 3.9. If the number of regeneration steps $i$ is very large, the initial encoded message from the redundant source is likely to have been modified and even maybe 'forgotten,' insofar as regeneration errors likely occurred. As the number of regenerations $i$ increases, the current encoded message thus depends less and less on the original redundant source, and more and more on the regeneration errors which occurred, i.e.,

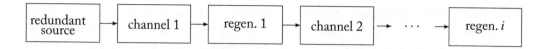

Figure 3.10: A chain of $i$ successive regenerations.

on a succession of contingent events. In the context of genetics, of course, natural selection always operates on the corresponding phenotypes, letting only the fittest ones survive. This illustrates the ambivalence of regeneration errors. On the one hand, they should occur very unfrequently for the sake of genome conservation, which is necessary for maintaining a species and multiplying the number of the individuals which belong to it. On the other hand, they give rise to new phenotypes when they occur, thus exploring at random the field of what is possible. These new phenotypes are targets of natural selection and the rough material which makes evolution possible. (The above statements are more precisely justified by computations in the analysis of the 'toy living world' introduced in Ch. 8.)

Remember that, as we noticed above, the roles of information and perturbations are relative to each other, entirely depending on their usefulness or harm for the destination in Shannon's paradigm. In genetics, the destination consists of the cell machinery which processes the genetic message borne by DNA. Therefore, the regeneration errors *create new genetic messages* in a very objective sense. There is no paradox in this statement since there is no difference of nature between information and perturbations. Because we liken the error-correcting codes to the multiple constraints which make a DNA string admissible as a genetic message (see Ch. 10 below), the DNA strings having suffered regeneration errors are just as plausible to the cell machinery as the error-free ones.

## 3.7    COMPUTING AN UPPER BOUND OF DNA CAPACITY

The ability of DNA to convey genetic information through time can be measured by its capacity, in the information-theoretic meaning of this word. To compute it, we need to know what kind of events impair the basic information-bearing elements, namely, the nucleic-base pairs; and at which frequency these 'error events' occur.

As regards the kind of error events which affect a nucleic-base pair, only four types need be distinguished: *substitution*: a wrong symbol replaces the correct one; *erasure*: the received symbol is recognized as foreign to the alphabet; *deletion*: a symbol is removed from the sequence; and *insertion*: a spurious symbol is inserted into it. We already gave as examples channels where substitutions and erasures occur: the binary symmetric channel and the binary erasure channel, respectively (Sec. 3.3.3; see Fig. 3.4). Error events of all four types may occur, but it is difficult to estimate in what proportion they do. Instead of precisely computing the DNA capacity, we can instead try to compute an *upper bound* on it. Information theory tells us that the mildest of all kind of errors is erasure[1] (in the above meaning) and we may, very optimistically, assume that only erasures occur.

How frequently should erasures be assumed to occur? Let the probability that an erasure occurs during the infinitesimal time interval d$t$ be denoted by $\nu$d$t$, where $\nu$ is referred to as the erasure frequency. We have no reliable estimate of $\nu$, which moreover may not be constant. (We will attempt in Sec. 7.3 to evaluate the frequency of substitutions, which seem to be the most current type of symbol error, using published biological literature and some assumptions.) For simply obtaining an upper bound on the DNA capacity, we may use a lower bound of $\nu$, which results in further overestimating the capacity.

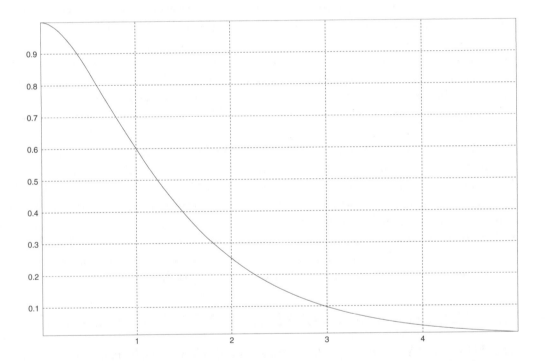

**Figure 3.11:** Upper bound on the DNA capacity as a function of time. The unit in the horizontal (time) axis is $1/\nu$.

The capacity of double-stranded DNA, assuming that only erasures occur at a constant frequency $\nu$, is computed in Ch. 7, and it is represented as a function of time in Fig. 3.11. The upper bound on the capacity thus obtained is an exponentially decreasing function of the product $\tau = \nu t$, where $t$ denotes time. This product can be interpreted as a measure of time using as unit the recip-

---

[1]Notice that the two words 'erasure' and 'deletion,' which are almost synonymous in the usual language, are given here very distinct meanings. Contrary to an erasure, the deletion of a symbol changes the sequence length from its initial value, say $n$, into $n - 1$.

rocal of the frequency, $1/\nu$. The frequency $\nu$ is actually unknown, but it cannot be extremely small, since mutations have deleterious effects at the time scale of individual lifetimes (genomes in somatic cells are presumably not strongly protected against errors, at variance with those of the germinal line). The time unit $1/\nu$ thus does not belong to the geological timescale.

We conclude from the fast decrease of the DNA capacity as time passes that mere copying does not suffice to ensure the faithful communication of genetic information. Genomes must be endowed with intrinsic error-correcting properties which enable their regeneration before the DNA capacity has become too low (remember that the capacity is an impassable limit). A genome must thus belong to an *error-correcting code*. An error-correcting code can reliably correct errors provided their cumulated number remains less than a threshold close to half its minimum distance. The conservation of a genome does not only imply that it belongs to an error-correcting code, but that the time interval between successive regenerations is as small as to keep the cumulated number of symbol errors below this threshold. The existence of an intrinsic genomic error-correcting code with a small enough time interval between successive regenerations is thus a mandatory condition of the faithful conservation of genomes, already referred to as our *main hypothesis*. A *subsidiary hypothesis* has been further introduced in order to account for the observed better conservation of very old parts of the genome (e.g., the *HOX* genes), which needs that the genomic code more than compensates the decrease of the DNA capacity. It assumes that the genomic error-correcting code actually consists of a combination of several nested component codes which have originated successively in time at the geological timescale. A new component code then does not only protect some new information against errors, but also provides further protection to the informations already encoded by the yet existing component codes. A layered structure results, where the information is the better conserved, the more central (the older) it is. Some uncoded information may be located in the most peripheral layer.

We already formulated the main and subsidiary hypotheses in Sec. 1.2.1 and briefly examined in the remainder of Sec. 1.2 their consequences on the living world and its evolution. These topics will be further developed in Ch. 11.

## 3.8    SUMMARY OF THE NEXT CHAPTERS

The remainder of this lecture is organized as follows. Part II is intended to complement the introduction as regards genetics and information theory. It begins with a rough overview of molecular genetics (Ch. 4). We then provide a short discussion of information theory, presented in a manner fitted to its application to genetics (Ch. 5), and then of error-correcting codes (Ch. 6). The third part establishes the need of genomic error-correcting codes and describes their impact on the living world. It begins with Ch. 7 where the information-theoretic capacity of a complementary pair of nucleotides is computed. We find that it vanishes exponentially fast as time passes, showing that DNA is an ephemeral memory at the timescale of geology. The faithful conservation of genetic information thus needs error-correction coding, our main hypothesis. A simple model referred to as 'toy living world' which illustrates this main hypothesis is introduced and quantitatively analyzed

(Ch. 8). It shows that only genomes involving error-correcting codes can be conserved at the geological timescale. We then state the subsidiary hypothesis that the genomic error-correcting code actually involves a multiplicity of component codes combined into a layered structure, referred to as *nested system*, in order to account for the better conservation of the older genomic information (Ch. 9). We introduce the concept of 'soft code', which weakens but widely extends that of error-correcting code, as plausibly describing how nature implements error correction (Ch. 10). From both the main and subsidiary hypotheses we then derive a number of consequences which are shown to properly fit actual features of the living world and of its evolution (Ch. 11). The actual mechanisms of genomic error-correction remain however to be discovered, a task which requires a close collaboration of engineers and biologists as discussed in Ch. 12 together with other open problems. Finally, Ch. 13 presents concluding remarks.

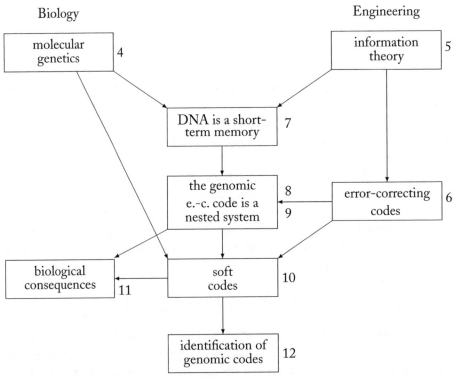

**Figure 3.12:** Connection between the topics of the lecture. The topic which labels a box needs as prerequisite that of the upstream boxes according to the arrows. The numbers near the boxes indicate the relevant chapter number.

Figure 3.12 schematically indicates the connection between the main topics of the lecture. It is intended to help the reader to find his/her own path in it, depending on his/her background and interest.

# Part II

# Facts of Genetics and Information Theory

CHAPTER 4

# More on Molecular Genetics

As with biology in general, molecular genetics is a complex topic. We shall deal very roughly with it: almost all statements in this survey, besides being simplified, actually have exceptions which are not mentioned for the sake of brevity and readability; similarly, only few references to the texts used for compiling this survey are given besides those of historical significance. We shall mention only the details of chemical structure which are relevant to our purpose.

## 4.1 MOLECULAR MEMORIES: DNA AND RNA

### 4.1.1 UNIDIMENSIONAL POLYMERS AS HEREDITARY MEMORIES

Since the works of Avery et al. [2] and Chargaff [26], it is known that the hereditary information is borne by *DeoxyriboNucleic Acid* (DNA) molecules in all living beings (except for certain viruses where ribonucleic acid, RNA, performs this role). DNA has been shown by Franklin and Gosling [38] and Watson and Crick [76] to assume the famous double-helix shape.

To convey information, a molecule must be a unidimensional polymer, i.e., a chain of an arbitrary number of links, each bearing a small molecule from a given finite repertoire, or alphabet. These information-bearing molecules succeed to each other in an arbitrary order; their sequence constitutes the genetic message. Similarly, a written text is a unidimensional sequence of letters, digits, separating spaces, and punctuation marks, i.e., symbols from a given alphabet, in the order that the author of the text chose. Unidimensionality is necessary in both a genetic message and a written text, since an order must exist between its symbols (information-bearing molecules or letters): given any two symbols, one of them should either precede or follow the other one. This is necessary for establishing syntactic rules, which are themselves necessary for the written message to convey a semantic content. No such 'relation of order' exists in two-dimensional space or in spaces having a larger number of dimensions.

### 4.1.2 STRUCTURE OF DOUBLE-STRAND DNA

The DNA molecule consists of two unidimensional chains (or strands) each made of alternating a phosphate group and deoxyribose (a pentose, i.e., a sugar having 5 carbon atoms) where each sugar molecule is covalently bound to one of 4 nitrogenous bases, referred to as nucleic bases or nucleotides, namely adenine (**A**), thymine (**T**), guanine (**G**), or cytosine (**C**). The nucleic bases in the two strands are such that **A** in one strand always faces **T**, and **G** in one strand always faces **C**, thus appearing in the DNA molecule as *complementary pairs*. The nucleic bases **A** and **G** are purines (abbreviated to **R**), i.e., two-cycle molecules, while **T** and **C** are pyrimidines (**Y**), single-cycle molecules. The bases **A** and **T** bound to facing strands are tied together by two hydrogen bonds, while three hydrogen bonds tie together **G** and **C**. A–T and G–C are referred to as complementary pairs, each made of a purine and a pyrimidine. It will be convenient to refer to these complementary pairs as weak (**W**) for **A–T** and strong (**S**) for **G–C**. The atoms in each complementary pair are contained in a same plane. The chemical formula of both nucleotide pairs is shown in Fig. 4.1.

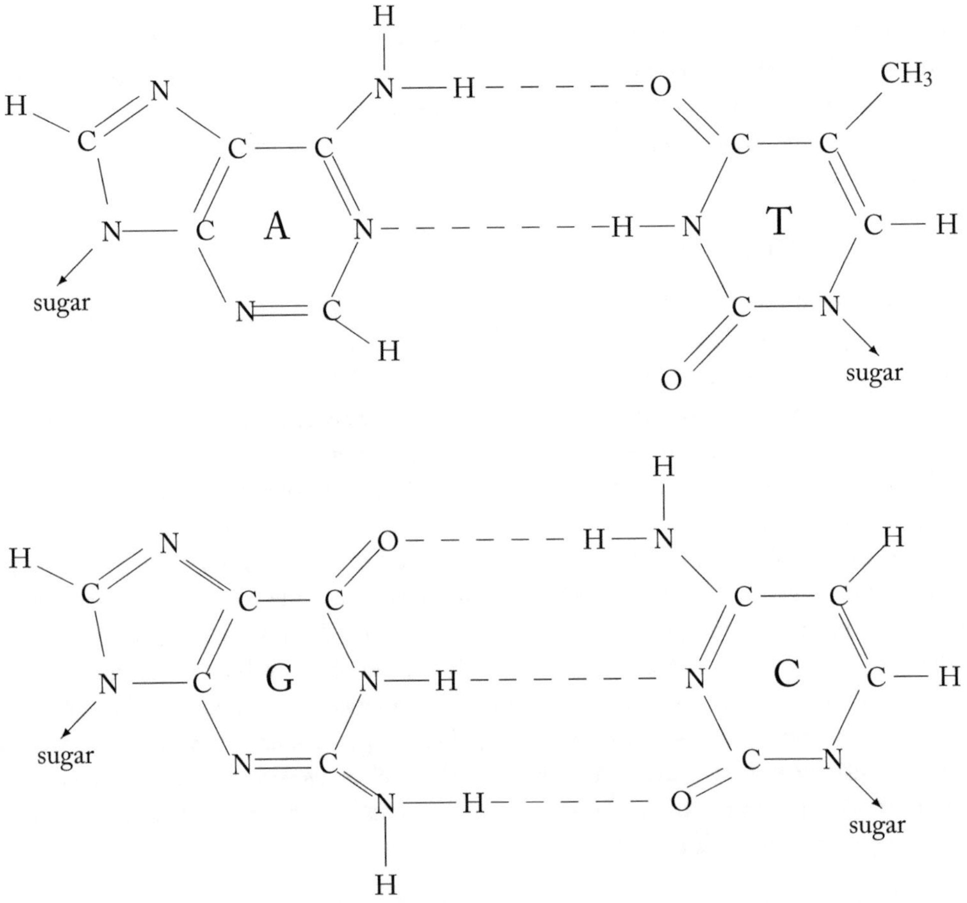

**Figure 4.1:** Chemical structure of the complementary base pairs **A–T** (top) and **G–C** (bottom) of DNA. The purines (**A** and **G**) are at left and the pyrimidines (**T** and **C**) at right. In RNA, the uracil molecule (**U**) replaces thymine (**T**), which is identical to it except that the methyl group $CH_3$ at top right is replaced by a hydrogen atom H. Dashed lines represent hydrogen bonds. The mutual distances of the atoms are roughly respected, showing that the two complementary base pairs have almost equal lengths.

The spatial length of the two complementary pairs is almost the same, which results in maintaining a constant distance between the two strands, and the angles of the covalent bonds with the sugar molecules of the backbones are also almost the same. Any complementary base pair can thus take place everywhere in the double-helix structure. The whole DNA molecule appears as a twisted ladder (a right double-helix) where the uprights are the two phosphate-sugar chains (also referred to as 'backbones') and the rungs are the pairs of complementary nucleic bases, which are located in a plane perpendicular to the axis of the double-helix. The width of the ladder is of about 2 nanometers, and the distance between two rungs of about 0.34 nanometers. A full turn of the double-helix corresponds to approximately 10 base pairs. A schematic ladder representation of the double-strand DNA molecule (ignoring the twist) is shown in Fig. 4.2.

The genetic information of a living being is borne by the sequence of nucleotides along one of the two strands of the DNA contained in its cell(s). Just like a written text is read in a direction defined by a convention (e.g., from left to right for texts in the Latin alphabet), the DNA molecule is read in the 5' to 3' direction, referring to the conventional numbering of the carbon atoms of the sugar molecule. Since the atoms of the backbones in complementary pairs are symmetric with respect to the center, the reading direction in one of the two strands of DNA is the inverse of that of the other strand.

## 4.1.3   RNA AS ANOTHER MOLECULAR MEMORY

The RNA molecule, very similar to DNA, is made of a 'backbone' to which nucleic bases are attached. Instead of deoxyribose, the sugar in RNA is the ribose, another pentose which differs from it by the presence of one more oxygen atom, and the nucleic base thymine (**T**) of DNA is replaced in RNA by uracil (**U**), a pyrimidine which only differs from thymine by a hydrogen atom replacing the methyl group (see Fig. 4.1). As **T**, **U** can be bound to **A** in a complementary pair. Double-strand RNA thus exists, although it is less stable than double-strand DNA. The main genetic memory in certain viruses is made of RNA. Except for these viruses, the main genetic memory of all living beings is made of DNA. RNA is more versatile: it is mainly found as an auxiliary memory in very many instances, especially in the cell replication machinery (see below). RNA has also some catalytic activity. An 'RNA world,' where RNA was acting both as a memory (like modern DNA) and as a catalyst (like modern proteins) is hypothesized to have predated the extant living world. Due to its double competence, RNA could catalyze its own replication.

## 4.1.4   DNA AS A LONG-LASTING SUPPORT OF INFORMATION

Regardless of its chemical and structural details, we may think of DNA as bearing the genetic message as a sequence of symbols from the quaternary alphabet {**A, T, G, C**}. Then heredity consists of communicating the genomic message through time, from the past to the present. We may thus deal with it as a problem of communication engineering. The DNA molecule is subject to errors due to chemical reactants and radiations, and it is moreover a quantum object in two of its dimensions, hence its deterministic description is impossible. The time scale involved is that of geology (remember that life originated on Earth about 3.5 billion years ago, and maybe even earlier), which makes communicating the genomic message a formidable problem. It turns out that its difficulty has been overlooked by biologists, who moreover failed to understand that information theory is

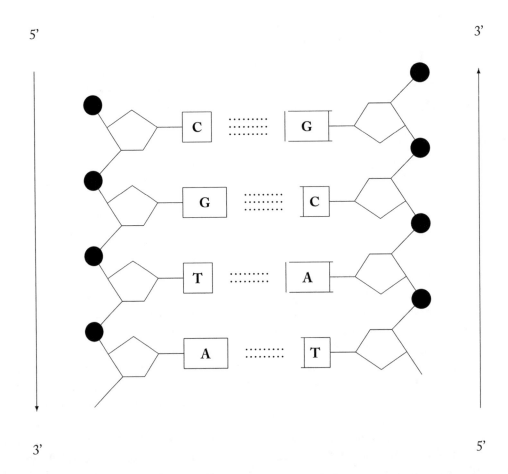

**Figure 4.2:** Ladder representation of double-strand DNA. The black disks represent phosphate groups and the pentagons, deoxyribose molecules (sugars). The polymers made of alternating phosphate groups and sugars are referred to in the text as 'backbones.' Rectangular boxes represent the nucleotides **A**, **T**, **G** and **C**. The dotted horizontal lines between nucleotides of opposite strands represent hydrogen bonds. The vertical arrows indicate the direction of reading.

relevant to it. Most of them believe that DNA replication according to the template-replication paradigm, i.e., splitting it into its two strands and appending to each of them its complementary sequence, suffices to faithfully communicate the genome through geological times (e.g., [29, 30]). Many DNA repair mechanisms are known to biologists. Their description is too often inaccurate in terms of communication engineering and the parameter values and performance measures given in the biological literature are doubtful. Taken together, these repair mechanisms succeed with an astonishing accuracy. It must be emphasized, however, that they concern means to avoid errors occurring *within the replication process*, a problem entirely different from making the genome resilient to casual errors: clearly, the faithful copy of a wrong original is wrong.

At the *geological time scale*, it is known that some parts of the genome (i.e., the *HOX* genes) are conserved with *great faithfulness*. Errors in DNA nevertheless occur, and moreover have an essential role in biological evolution. They are referred to as *mutations*. They occur in somatic cells at the much shorter time scale of individual life times. The blatant contradiction between the long-term faithful conservation of the genomes and the existence of mutations at a much shorter time scale has been our main motivation for undertaking the research reported in this book. The faithful communication of a genome through time demands that it is made resilient to errors, which can only be obtained by the use of error-correcting codes which enable *regenerating* it, not merely copying it, however faithfully; see Sec. 1.3 above.

### 4.1.5    ERROR-CORRECTION CODING AS AN IMPLICIT HYPOTHESIS

The existence of error correction mechanisms is indeed implicit in the biological literature. Biologists consider as self-obvious the fact that the mutation rate depends on the species, the locus or the nature of the gene (for instance, the *HOX* genes control the expression of other genes and determine the organization plan; since they are shared by almost all multicellular living beings, they are considered as devoid of any mutation). Biologists also admit that the mutation rate of a gene can be controlled by factors like other genes. The lowest mutation rate reported is found in nonsynonymous mutations (i.e., substitutions of a single nucleotide which change an amino-acid into another; since the genetic 'code' is a many-to-one mapping, different codons may correspond to the same amino-acid), namely $10^{-9}$ substitutions per year and per nucleotide in the genome of mammals. In contrast, mutations of the same kind occur at a rate of $14 \times 10^{-3}$ substitutions per year and per nucleotide in a region of *env*, one of the 9 genes of the HIV virus [28]. The effects on the phenotype of a nonsynonymous mutation are highly variable. It sometimes destroys or changes the biological function of the protein, but has often a more limited effect.

Similarly, the repeated use of the rather convincing metaphor of a written text to explain the role of the genome by Dawkins [30] and many others implies the existence of genomic error-correcting codes. Since a language is subject to many constraints at several levels (morphological, grammatical, semantic, etc.), it may rightfully be interpreted as a kind of highly redundant error-correcting system. Interestingly, this several-level system resembles the nested system we assumed according to our subsidiary hypothesis. Moreover, we find the same kind of object (to be referred to as 'soft code' in Ch. 10) when trying to guess what the error-correcting codes we hypothesized look like.

## 4.2   PLACE AND FUNCTION OF DNA IN THE CELL

### 4.2.1   CHROMOSOMES AND GENOMES

The double-strand DNA sequences are combined with certain proteins (histones) and more or less closely packed in large aggregates named *chromosomes*. The number of chromosomes depends on the species, e.g., a bacterium has a single circular chromosome but the living beings higher in the evolutionary scale generally have several rod-shaped ones. For instance, the somatic (nonsexual) human cells contain $2 \times 23$ chromosomes. The sequence of base pairs in the whole set of chromosomes of a living being is referred to as its *genome* and the total number of base pairs it contains, as the genome length. The genome length varies from about a million base pairs for certain bacteria up to billions in eukaryotic beings. Genomes are thus extremely long (e.g., the total length of double-strand DNA in each somatic human cell, which contains two copies of the genome, is about two meters; much longer genomes moreover exist). Viruses generally have shorter genomes, typically of a few thousands base pairs and, as stated above, some of them are made of RNA. There is clearly a considerable variation in the genome length from a living being to another. The longest genomes are not related to the highest complexity beings: as the most striking example, the unicellular *Amoebia dubia* has the longest known genome, more than 200 times longer than the human one [73, 28, 47]. In *prokaryotes* like bacteria, the genome is not physically separated from the remainder of the cell, whereas in *eukaryotes* (all multicellular organisms and certain unicellular ones like yeasts or amiboeas) it is contained in the *nucleus*, a region separated from the remainder of the cell by a membrane. The eukaryotic cell is much larger and more complicated that the prokaryotic one and probably younger by almost 2 billion years. Being contained inside the cell (for prokaryotes) or the cell nucleus (for eukaryotes), which are both a few micrometers long, DNA is thus very densely packed, although variably along chromosomes: less densely in euchromatin than in heterochromatin. In the nucleus of eukaryotic cells, the double-stranded DNA is packed into nucleosomes, which consist of a histone octamer acting as a spool with almost two turns (about 165-nucleotide long) of double-stranded DNA wrapped around it. (Histones are proteins associated with DNA in chromosomes.) Moreover, nucleosomes themselves are packed together in higher order structures generally referred to as 'fibers.'

### 4.2.2   PRINCIPLE OF DNA REPLICATION

The whole genome can be faithfully reproduced. The hydrogen bonds which tie together the nuclotides of complementary pairs are much weaker than the covalent bonds which bind the nucleotides to the backbones as well as the atoms of the backbones themselves. Splitting the double-strand DNA into two single strands is thus easy (requires few energy), and the nucleotides of each single-strand DNA can be endowed with their complementary nucleotides so as to restore two double-strand DNA molecules, identical in principle to the initial double-strand one. This is the *template-replication paradigm*. In the transcription process (see below), one of the strands is used as a template to generate a messenger (or premessenger) RNA molecule which is a copy of the other strand, except that uracyl, **U**, replaces thymine, **T**.

### 4.2.3   GENES INSTRUCT THE SYNTHESIS OF PROTEINS

Besides its own replication, the genome also directs the construction of a *phenotype*. The best under-stood function in this respect is the synthesis of *proteins* which are the main constituents of living matter and actors of life processes. Whereas replication concerns all the DNA present in the cell in the case of a prokaryote, or its nucleus for a eukaryote, only specific regions of the DNA strands can direct the synthesis of proteins. They are named *genes*. Each of them is precisely located at some *locus* in a chromosome. Non-identical genes which can be found at a same locus in different individuals of the same species are referred to as *alleles*. Proteins and other genes inhibit or enable the synthesis of the protein for which a given gene 'codes' (its *expression*). In multicellular organisms, cells are highly specialized so the on/off switching of the gene expression at a given locus depends on the cell type and history, although each cell contains the whole genome.

The fraction of the genome actually made of genes, hence actually used for synthesizing proteins, is greatly variable, from about 100% in bacteria to 0.02% or less in beings like fritillary or lungfish which have a very long genome. It is of about 1.3% in the human genome [73, 28, 47, 4]. The number of genes varies from about 5,000 in bacteria to about 29,000 in humans (the exact figure is not yet precisely known, all the more since there is no agreement on the gene definition [47, 4]). There is thus much less variation in the number of genes than in the length of genomes of living beings.

The synthesis of a protein results from the gene directing the assembly of a sequence of amino-acids (the polypeptidic chain) to be later properly folded. This process is performed in two successive steps: 'transcription' and 'translation'. Transcription consists of copying the gene into an RNA string referred to as *messenger RNA* (mRNA). Translation then consists of reading the mRNA in an organelle named *ribosome* where triplets of successive nucleotides, referred to as 'codons,' determine amino-acids according to a rule which is referred to as the *genetic 'code'*[1]. Each codon gives rise to a short molecule of 'transfer RNA' which binds itself to the amino-acid which corresponds to this particular codon according to the genetic 'code.' Figure 4.3 is a chart of the genetic 'code.'

### 4.2.4   AMINO-ACIDS AND POLYPEPTIDIC CHAINS

An amino-acid is made of a carbon atom (referred to as $C_\alpha$) which is bound to a hydrogen atom, an amino group ($NH_2$), a carboxy group (COOH), and a specific side chain. The amino-acids which correspond to successive codons are linked together by binding the carboxy group of an amino-acid and the amino group of the other one, with elimination of a water molecule (see Fig. 4.4). The order of the amino-acids is thus determined by that of the codons in the messenger RNA. A 'polypeptidic chain' results. It becomes a protein after its proper folding.

Table 4.1 lists the 20 amino-acids which constitute polypeptidic chains. They differ from each other as regards their side chain. Depending on the properties of these side chains, the amino-acids can be put into four main classes: hydrophobic (abbreviated as h-phobic), hydrophilic (h-philic), acid

---

[1]In the vocabulary of information theory, it is an encoding rule rather than a code *stricto sensu*. We are led to assume the existence of a *genomic code* in a quite different meaning, so when referring to it we always use quotes.

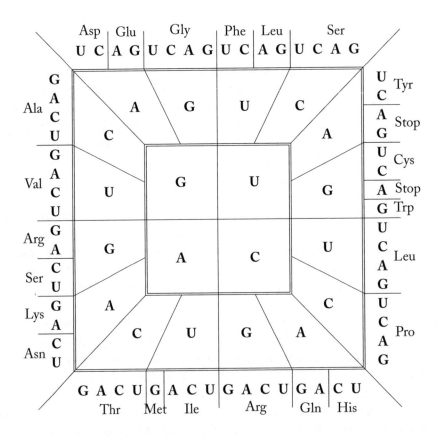

**Figure 4.3:** The genetic 'code'. This chart indicates how the codons of messenger RNA specify amino-acids or stop the synthesis of a polypeptidic chain. The letters **A, C, G,** and **U** denote the nucleic bases of RNA. Any codon is a three-nucleic-base word read from center to periphery. Each three-letter acronym at the periphery indicates to which amino-acid a codon corresponds, if it is not a stop command. For instance, **UAC** 'codes' for the amino-acid tyrosine (Tyr), while **UAA** stops the translation process.

(bearing a negative electric charge), and base (bearing a positive charge). Except for glycine which has a single hydrogen atom as side chain, they all are chiral molecules (i.e., they can assume two forms which are symmetric with respect to a plane in the three-dimensional space, like an object and its image in a mirror), but only their L-form occurs in proteins. The reason why the amino-acids listed above were 'chosen' rather than others is unknown, as well as why the L-form has been 'preferred.' Cystein, whose side chain is the group SH, can form disulphur bridges with another cystein molecule close to it in the three-dimensional space due to the protein folding.

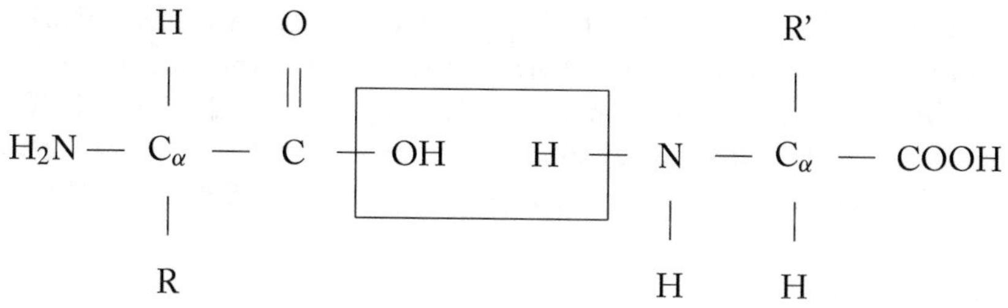

$$H_2N - C_\alpha - C + OH \quad H + N - C_\alpha - COOH$$

with H, O above ($H$ over $C_\alpha$, $O$ double-bonded to $C$), R below $C_\alpha$, and R' above the second $C_\alpha$, H and H below N and second $C_\alpha$.

**Figure 4.4:** Two amino-acids and their linking into a polypeptidic chain. R and R' denote their side chains. Their linking results from eliminating the water molecule shown inside the box.

| acr. | name | class | # | acr. | name | class | # |
|------|------|-------|---|------|------|-------|---|
| Met | methionine | h-phobic | 1 | Tyr | tyrosine | h-philic | 2 |
| Trp | triptophan | h-phobic | 1 | Ile | isoleucine | h-phobic | 3 |
| Asp | aspartic acid | acid | 2 | Ala | alanine | h-phobic | 4 |
| Glu | glutamic acid | acid | 2 | Gly | glycine | h-phobic | 4 |
| Asn | asparagine | h-philic | 2 | Pro | proline | h-phobic | 4 |
| Cys | cysteine | h-phobic | 2 | Thr | threonine | h-philic | 4 |
| Phe | phenylalanine | h-phobic | 2 | Val | valine | h-phobic | 4 |
| Gln | glutamine | h-philic | 2 | Arg | arginine | base | 6 |
| His | histidine | base | 2 | Leu | leucine | h-phobic | 6 |
| Lys | lysine | base | 2 | Ser | serine | h-philic | 6 |

Table 4.1: Amino-acids which constitute polypeptidic chains

## 4.2.5 SYNTHESIS OF A POLYPEPTIDIC CHAIN

In eukaryotic genes, DNA transcription is not directly followed by translation. It results in a 'premessenger RNA' which has to be further transformed before it directs the synthesis of a polypeptidic chain. Some parts of it are cutout and the remaining parts are attached together, resulting in the mature messenger RNA. This intermediate step is named 'splicing.' The removed parts (from the premessenger RNA, or the corresponding parts of DNA) are referred to as 'introns,' and the remaining ones as 'exons.' Only exons thus contribute to the synthesis of proteins. A further complication results from the possible rearrangement of the spliced segments of RNA according to several distinct patterns, resulting in possibly different polypeptidic chains, hence in different proteins. The rule formerly believed to be true 'a gene, a protein' is then violated and a single gene then generates several proteins.

The process of transcription is performed by an enzyme which reads one of the DNA strands in a well defined direction (5' to 3'), just like a reading head reads a magnetic tape. Translation similarly proceeds in the same direction on the messenger RNA (the 'reading head' is then the ribosome). Particular sequences referred to as 'promoters' are first recognized which warn that the translation will have to take place farther. Then, it actually starts when the particular codon **AUG**, which 'codes' for the amino-acid methionine, is met. The position of the codon **AUG** also indicates the 'reading frame,' i.e., how the mRNA strand has to be segmented into three-nucleotide codons. The translation continues until one of the three stop codons **CAA**, **CAG**, or **CGA** is found.

### 4.2.6    PROTEINS

A protein results from the proper folding of a polypeptidic chain, which results itself from the translation of a messenger RNA into a sequence of amino-acids. Typically, a protein is made of a few hundreds of amino-acids, each of them being one of the 20 amino-acids listed in Tab. 4.1.

A protein results from a *polypeptidic chain* by appropriate foldings which give it a unique three-dimensional shape determined by chemical affinities of the side chains of its constituents, especially hydrogen bonds and disulphur bridges, which result in establishing bonds between amino-acids at a distance along the polypeptidic chain. The biological properties of a protein heavily depend on its spatial structure (e.g., anomalous folding of the prion is responsible for the BSE disease). A protein combines several substructures like $\alpha$-helices and $\beta$-sheets, which themselves are combined into higher order structures named 'domains.'

Proteins are the most important constituents of the living matter, having both a structural and catalytic role. When they act as catalysts, they are referred to as *enzymes*. DNA replication, especially, as well as DNA transcription and translation, are catalyzed by proteins. Therefore, proteins catalyze their own synthesis. Most enzymes have a roughly globular shape with an 'active site' where the catalyzed reaction takes place.

## 4.3    GENOME AND PHENOTYPE

### 4.3.1    A GENOME INSTRUCTS THE DEVELOPMENT AND MAINTENANCE OF A PHENOTYPE

The sequence of protein syntheses in a reproductive cell like a fertilized egg results, if the environment is favourable enough, in the development of a *phenotype* through very complex processes which are still poorly understood. During these processes the genes are expressed at the appropriate place and time.

The phenotype hosts and protects the genome. Both together constitute a living being. The ability of the genome to reproduce itself implies that an initial DNA chain can generate other ones, identical to it in the absence of replication errors, which will give rise to other phenotypes. An initial genome can thus generate a population of ideally identical living beings (except for replication errors), its children, which themselves generate grandchildren, etc. Any descendant ideally bears the same information as the initial genome although it was born later so we may think of this property as *communication through time*. The hereditary features of living beings are solely determined by the information borne by the genome thus communicated through time. There is evidence that the origin of life has been a single event so the present beings bear (as genomes), and are made according to (as phenotypes), an information which dates back to geological times. Remember that the age of the Earth is about 4.6 $10^9$ years and the very beginning of life probably occurred about 3.5 $10^9$ years ago and maybe earlier.

### 4.3.2    A PHENOTYPE HOSTS THE GENOME FROM WHICH IT ORIGINATES

A phenotype contains the genome from which it originates and provides the environment and machinery necessary to its maintenance and replication. Its membranes shield it against mechanical and chemical aggressions. The genome cannot be conserved outside the phenotype. In prokaryotes like bacteria, the genome is contained inside the cell, which is separated from the outside by a single membrane. In eukaryotes, a second membrane delimitates inside the cell a nucleus which contains the genome, which is thus separated from the outside of the cell by two membranes. Membranes do not perfectly shield genomes against damages of all kinds hence errors may occur. Errors may also result from inaccuracies in the replication process, although many 'proof-reading' mechanisms almost always result in a faithful copy, but even a perfect replication faithfully reproduces the changes suffered by the genome itself hence does not prevent errors of other kind.

According to the *central dogma of molecular biology*, the role of memory is exclusively devoted to DNA. It states that information can never go from proteins to DNA although the polypeptidic chain from which a protein originates, too, is a sequence of elements belonging to a set of a few molecules. Not in contradiction with this 'dogma,' there exists a process of *reverse transcription* such that DNA can be synthesized from RNA. It is believed to have generated a sizeable portion of the genome in many organisms, including the human one [4]. It appears as a tool for lengthening the genome by repeating some parts of it. The central dogma excludes any influence of the phenotype on the genome so it rules out the Lamarckian type of evolution according to which characters acquired by phenotypes could be inherited. Evolution can thus only be Darwinian i.e., genomes suffer variations, resulting in a number of phenotypes which are the target of natural selection. The only remaining genomes are those associated with the surviving phenotypes.

## 4.4    DNA RECOMBINATION AND CROSSING OVER

In higher beings with sexual reproduction, e.g., in mammals, the somatic cells are diploid, i.e., each one contains two homologous copies of each chromosome, inherited from each of their parents, at variance with their sexual cells (gametes) which are haploid, i.e., contain a single copy of each of the chromosomes. The process of gamete formation, referred to as *meiosis*, thus implies that a diploid cell with two copies of each of its chromosomes gives birth to two gametes having each a single copy of each chromosome. This single copy is actually a mixture of the two parental copies, made of alternating segments from either the paternal or the maternal copy. The choice of these segments seems to be random. The process which assembles these segments into a new chromosome is referred to as *crossing-over*. The alleles of both parental copies of each chromosome are thus combined into a new sequence (remember that alleles are different alternative versions of a same gene). Fertilization will eventually restore a diploid cell, the fertilized egg, with two copies coming from each of the gametes, each of them being itself such a mixture of segments inherited from two parents. Meiosis and fertilization together result in generating new genomes by combining alleles. Each of these genomes may be considered as unique because the number of possible combinations of alleles is immense.

The ability of double-strand DNA to recombine is an important property which is not limited to the crossing-over which takes place at meiosis. If two double-strand DNA strings bear close enough sequences of base pairs, they can similarly exchange segments of genetic material when put together (see Ch. 14 in [37]).

CHAPTER 5

# More on Information Theory

This chapter gathers sections intended to complement the overview of information theory presented in Ch. 3. Examples will often be given using the binary alphabet. Remember that we use the word 'bit,' as an abbreviation of 'binary digit,' in order to designate the symbols of the binary alphabet and never as meaning the binary information unit which we name 'shannon' (see Sec. 3.3.1).

## 5.1 ALPHABET, SOURCES, AND ENTROPY

An information is communicated by causing the occurrence of an event belonging to some predetermined set of $M$ possible events which can be recognized from each other. According to the quantitative measure of information introduced in Sec. 3.3.1, a single occurrence brings an information quantity of at most $\log_2 M$, a finite quantity regardless how large is $M$. The solution for communicating an unlimited quantity of information consists of causing the occurrence of *successive* events, the number of which is unlimited. If the information-generating event is the choice of a particular symbol among an alphabet of size $q$, $M = q^n$ possible distinct messages result from $n$ successive such choices, providing the quantity of information $\log_2 M = n \log_2 q$ shannons which is unlimited as proportional to the natural number $n$. An unlimited information quantity of $n$ shannons is obtained even for the smallest possible alphabet size $q = 2$. Moreover, the necessary condition that the events be recognizable is more easily fulfilled for small values of $q$. Notice that the symbol choices need be successive, which implies their ordering, hence that they occur within a unidimensional space. It is so for ordinary communication (from a point of space to another one) since in this case the information bearing-events successively occur in time. Using more than one dimension would make this ordering impossible. As a matter of fact, heredity as well as the written human language use unidimensional sequences although they could conceivably use the available three spatial dimensions of physics in order to communicate through time. Moreover, the very existence of grammatical rules and semantics needs ordering of symbols in the case of the human language and, very probably, of DNA-borne heredity.

### 5.1.1 MEMORYLESS SOURCES, MARKOVIAN SOURCES, AND THEIR ENTROPY

We introduced in Ch. 3 a simple example of Markovian source. We now give a more comprehensive definition of such a source.

Figure 5.1 represents two types of sources: a memoryless source (Fig. 5.1-**a**) and a Markovian source (Fig. 5.1-**b**). A $q$-ary memoryless source chooses repeatedly (for instance, at each pulse of a periodic clock) a symbol from an alphabet of size $q$ with constant probabilities $p_1, p_2, \ldots, p_q, p_i$

being the probability that $i$ is chosen (without loss of generality, we denote the alphabet symbols by $1, 2, \ldots, q$); we assumed $q = 2$ in the figure and we denoted the probability of choosing 1 by $p$, so that of choosing 0 is $1 - p$). These probabilities sum up to 1, i.e., $\sum_{i=1}^{q} p_i = 1$, and are assumed not to vary with respect to time, so this source is said to be *stationary*. The information quantity associated with the choice of symbol $i$ is $h(i) = -\log_2(p_i)$ shannons, so the average amount of information associated with the source is the mean (or expectation) of this quantity, namely

$$H = -p_1 \log_2(p_1) - p_2 \log_2(p_2) - \ldots - p_q \log_2(p_q) = -\sum_{i=1}^{q} p_i \log_2(p_i), \qquad (5.1)$$

referred to as the entropy of the source. It is well defined for a stationary source. For historical reasons, Shannon chose the name 'entropy' which was already that of a fundamental thermodynamic entity, but the relation between both is not straightforward and will not be discussed here. The information $h(i) = -\log_2(p_i)$ is a positive quantity since a probability is less than or equal to 1, hence its logarithm is negative. $H$ is thus positive except if one of the outcomes has probability 1; in this case $H = 0$ so, in more rigorous terms, $H$ is actually nonnegative. It is easy to show that the maximum $H_{\max}$ of $H$ is obtained when $p_1 = p_2, = \ldots = p_q = 1/q$ and thus equals $\log_2(q)$ shannons. The difference $r = H_{\max} - H$ between this maximum and the actual source entropy measures the *redundancy* of the source.

A Markovian source has been represented in Fig. 5.1-**b**. It combines two elements: a memoryless source and an $m$-symbol memory. This memory records the last $m$ symbols generated by the source and can thus assume $q^m$ distinct states. The probabilities of symbol choice of the associated memoryless source now depend on the state $s$ of the memory: symbol $i$ is chosen with probability $p_i(s)$. This scheme can generate much more complex sequences than a memoryless source since they depend on the $q^m$ probability distributions on the alphabet symbols associated with the memory state, instead of a single one. The probabilities $p_i(s)$ are constant for all possible values of $i$ and $s$ so the source is again stationary. The above definition of the entropy needs to be extended. The entropy of the source when it is in state $s$ is expressed according to Eq. (5.1) as:

$$H(s) = -p_1(s) \log_2[p_1(s)] - p_2(s) \log_2[p_2(s)] - \ldots = -\sum_{i=1}^{q} p_i(s) \log_2[p_i(s)]$$

and the overall entropy of the source is defined as the average of this quantity with respect to all possible states, namely:

$$H = \sum_{s=0}^{q^m - 1} \Pr(s) H(s), \qquad (5.2)$$

where $\Pr(s)$ is the steady probability of state $s$ and $H(s)$ has just been defined. Notice that the entropy defined by Eqs. (5.1) or (5.2) is an entropy *per symbol*. The entropy of a sequence of length $n$, which measures the average information it bears, is thus $nH$.

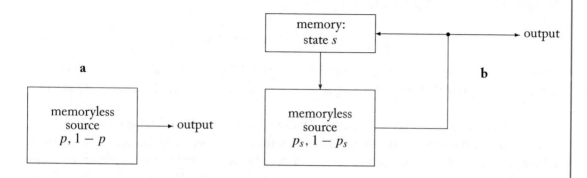

**Figure 5.1:** Two examples of binary sources. **a**: memoryless source. Its output is 1 with probability $p$, 0 with probability $1 - p$. **b**: Markovian source. The last $m$ output bits are recorded in the box labeled 'memory,' and the memoryless source below (identical to **a**) outputs 1 with probability $p_s$ (0 with probability $1 - p_s$) where probability $p_s$ depends on the state $s$, $s = 0, 1, \ldots, 2^m - 1$, of the memory.

A detailed example of Markovian source is given in Appendix 5.A below. Less simple sources can be defined, but the memoryless and Markovian sources will be sufficient for us to understand the main properties of sources in general.

The concept of Markovian source is also useful to illustrate the important notion of *ergodicity*. If all the probabilities $p_i(s)$ are different from 0 or 1, each of the possible states of the memory is sooner or later visited and observing the generated sequence enables in principle to assess the frequency of occurrence of each state, meaning that accurate estimates of the probabilities $p_i(s)$ can be obtained when the observation time is long enough. This may be thought of as a consequence of the (weak) law of large numbers. The source is then said to be *ergodic*: the indefinitely lengthened observation of the source output provides sufficient knowledge of it. Let us assume on the contrary that one of the $p_i(s)$ equals 0 or 1, for instance $p_0(0) = 1$. Then, once the 0-state is reached, the source only transmits zeros. The uninformative all-0 sequence is then generated, and the existence of states other than the 0-state cannot be deduced from observing the generated sequence. In other words, the all-0 state acts as a trap which hinders the source evolution towards other states. Such a source is nonergodic. *A contrario*, the beginning of the operation of an ergodic source, which implies some initial state, is assumed to have occurred in an arbitrarily remote past, making the observed source operation independent of the initial memory content. In general, a Markovian source is ergodic if there is an uninterrupted path between any two of its states, 'uninterrupted' meaning that no transition between the states of the path has zero probability. Although stationarity and ergodicity are often associated, there is no relationship between these two concepts.

For more general sources (i.e., which are neither memoryless nor Markovian), the entropy can be defined as the following limit, for $n$ approaching infinity:

$$H = - \lim_{n \to \infty} \frac{1}{n} \sum Pr(\underline{s}) \log_2[Pr(\underline{s})] \tag{5.3}$$

where $\underline{s}$ denotes a sequence of length $n$ and $Pr(\underline{s})$ its probability; the sum is extended to all possible sequences of length $n$ that the source can generate. This limit exists provided the source is stationary.

## 5.1.2   A FUNDAMENTAL PROPERTY OF STATIONARY ERGODIC SOURCES

Let us consider the set of all sequences of length $n$ that a stationary ergodic source can generate, $n$ being an arbitrarily large integer. According to the *Shannon-McMillan theorem*, these sequences can be divided into two mutually exclusive groups: that of *typical sequences*, each of which occurring with a probability close in some sense to $2^{-nH}$, where $H$ is the entropy as defined by Eq. (5.2); and the sequences which are not typical and thus referred to as *atypical*. The theorem states that the probability that an atypical sequence occurs vanishes as the sequence length $n$ approaches infinity. Only typical sequences thus occur with a nonnegligible probability, which is said for short *almost always*. This result is far from trivial, since the total number of typical sequences is of about $2^{nH}$, which shows that they are a minority among all the $q^n = 2^{n \log_2(q)} = 2^{nH_{max}}$ *a priori* possible sequences for an alphabet of size $q$. The last equality is true because the maximum possible value of the entropy, for this alphabet, is $H_{max} = \log_2(q)$. The ratio of the number of typical sequences to that of all possible ones is $2^{n(H-H_{max})}$, or $1/2^{nr} = 1/(2^n)^r$, if we define the redundancy $r$ as $r = H_{max} - H$. The redundancy then appears as measuring how few are the typical sequences with respect to all possible ones. The proof of this theorem is rather easy for Markovian sources, but it involves somewhat lengthy calculations, so we omit it here. It can be found in any textbook on information theory, e.g., [27]. Moreover, this theorem holds true for any stationary ergodic source, whether it is Markovian or not.

Let us compare a redundant source as just described with an error-correcting code. We found that the sequences of length $n$ generated by the source can be divided into two mutually exclusive groups: that of typical and atypical sequences. Almost all generated sequences belong to the first group, although they are only a minority among the possible $q^n$ sequences of length $n$. We also defined an error-correcting code of length $n$ as a strict subset of sequences, say of $q^k$ sequences with $k < n$, among the possible $q^n$ sequences of length $n$. The two previous sentences have the same meaning except that using the word 'almost' in the first one expresses a reservation which becomes less and less relevant as the sequence length increases. The set of sequences generated by a redundant source is thus clearly similar to an error-correcting code, which may be thought of as an information-theoretic justification of the concept of 'soft code' to be introduced in Ch. 10. The major difference is that the mutual Hamming distances between the sequences generated by a redundant source may not be favorable to error correction as not being designed for this purpose, but codes chosen at random are known to be good error-correcting codes (see Sec. 3.5.1). Then soft codes defined by constraints not intended to error correction are nevertheless likely to be good for this purpose[1]. Moreover, the soft codes of biology result from an evolutive process: they have been subjected to natural selection so it is likely that those which survived are good.

## 5.2   ABOUT SOURCE CODING

Although it is unlikely that source coding is directly implemented in genomes, it is interesting in order to simply introduce some important information-theoretic concepts. They will especially be useful for understanding some aspects of the algorithmic information theory (see Sec. 5.4).

### 5.2.1   SOURCE CODING USING A SOURCE EXTENSION

The $k$-th extension of a $q$-ary source $S$, denoted by $S^k$, is the source which results from grouping the successive symbols of its output messages by blocks of $k$ symbols and dealing with them as symbols of a $q^k$-element alphabet. Notice that the $k$-th extension $S^k$ of a source $S$ is an alternative way of describing it, not a different source.

For instance, let a memoryless binary source generate the message

$$000\ 100\ 101\ 101\ 011\ 100\ 110 \ldots$$

The corresponding sequence generated by its third extension is

$$\underline{0}\ \underline{4}\ \underline{5}\ \underline{5}\ \underline{3}\ \underline{4}\ \underline{6} \ldots,$$

the symbols of which now belong to the 8-element alphabet $\{\underline{0}, \underline{1}, \ldots, \underline{7}\}$. (They are denoted by underlined digits in order to distinguish them from the symbols of the original source.) Each of its symbols is represented by the corresponding block in natural binary numeration. Let us assume for instance that a memoryless binary source, say $S$, generates the symbol '0' with probability $p_0$ and symbol '1' with probability $p_1 = 1 - p_0$. Then, its third extension $S^3$ generates symbols $\underline{0}$ to $\underline{7}$ with probabilities $p_0^{k-w} p_1^w$ where $w$ denotes the weight of the corresponding 3-bit block of $S$, i.e., the number of ones it contains (here and in the sequel we use the word 'bit' as an acronym for 'binary digit'). For instance, if $p_0 = 0.9$ in $S$, the probabilities associated with the symbols of $S^3$ are 0.729 for $\underline{0}$, 0.081 for $\underline{1}, \underline{2}$ and $\underline{4}$, 0.009 for $\underline{3}, \underline{5}$ and $\underline{6}$, and 0.001 for $\underline{7}$.

The $k$-th extension of a source can be used in order to perform source coding. It basically consists of replacing the blocks of the $k$-th extension by *words* the shorter, the most probable is the corresponding block. A *variable length code* results. Using binary words, we may for instance encode the third extension $S^3$ defined above according to the code defined by the following table. Then a single-bit word is associated with the most probable symbol of the 3-rd extension, which in the above example is $\underline{0}$, and words with other symbols the longer, the less probable they are. Using this code results in an average length of 1.598 per triplet of the original source, meaning that the length of the message is reduced by the factor $1.598/3 = 0.53266 \ldots$

---

[1]This sounds paradoxical, but the difficulty with randomly chosen codes is the prohibitive complexity of their decoding. The soft codes can be useful for error correction if the constraints they obey favor their easy decoding.

| symbol | codeword | prob. |
|--------|----------|-------|
| 0 | 0 | 0.729 |
| 1 | 100 | 0.081 |
| 2 | 101 | 0.081 |
| 3 | 11100 | 0.009 |
| 4 | 110 | 0.081 |
| 5 | 11101 | 0.009 |
| 6 | 11110 | 0.009 |
| 7 | 11111 | 0.001 |

Table 5.1: A variable length code for source coding of $\mathcal{S}^3$ as defined in the text

To be useful, a variable length code must enable uniquely separating the codewords, i.e., recognizing the beginning and the end of each of them when they are transmitted successively. A variable length code endowed with this property is said *uniquely decipherable*. Several means to design a code with this property exist, including the trivial use of a symbol exclusively intended to mean the word separation (as the space used in written human language). An especially interesting family of uniquely decipherable codes are the *instantaneous codes*, also referred to as *irreducible*. (The word 'instantaneous' means that a sequence which is a word of the code is recognized as such as soon as it has been received.) These codes obey the constraint that no codeword begins with a sequence which is another codeword, referred to as the 'prefix condition' for no codeword may be a prefix to another one. One easily checks that this condition is actually fulfilled in the code given as an example in Tab. 5.1. This code was actually designed using the Huffman algorithm, which results in an instantaneous code having the minimum possible average length [46].

## 5.2.2   KRAFT-MCMILLAN INEQUALITY

Let an instantaneous code consist of $N$ codewords of respective lengths $\ell_1, \ell_2, \ldots, \ell_N$ (in nondecreasing order) and $q$ denote the alphabet size. This code then obeys the Kraft inequality

$$\sum_{i=1}^{N} q^{-\ell_i} \leq 1 \,. \tag{5.4}$$

Conversely, if $N$ integers $\ell_1, \ell_2, \ldots, \ell_N$ obey (5.4), an instantaneous code exists which has these integers as word lengths.

This inequality can be given a very simple proof. It relies on a graphical representation of all the $q$-ary words of length $\ell_N$ by means of a *tree*. Such a tree is depicted in Fig. 5.2. We assumed for drawing it $q = 2$ and $\ell_N = 4$; $q$ branches diverge from a point referred to as the root of the tree. The other extremities of these branches are referred to as the 1-st level nodes. Then $q$ branches diverge

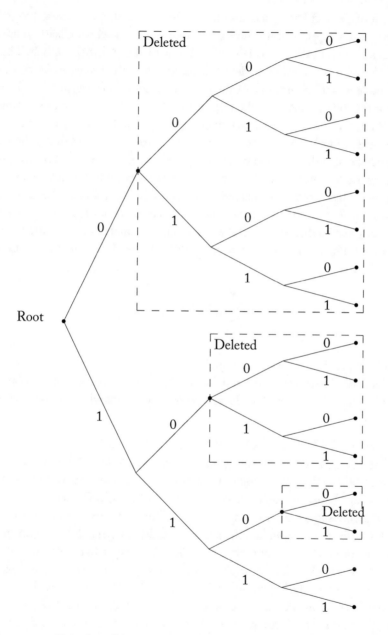

**Figure 5.2:** Binary tree of length 4. An ascending branch represents the bit 0 and a descending one the bit 1. Deletion of the partial trees as indicated in the figure results in the tree representing the five-word irreducible code {0,10,110,1110,1111}.

again from each of the 1-st level nodes and their other extremities are referred to as the 2-nd level nodes. This process is repeated until the $\ell_N$-th node level is reached. Clearly there are $q^{\ell_i}$ nodes of level $\ell_i$. Each branch in the tree is labeled with one of the alphabet symbols. The $q^{\ell_N}$ nodes of level $\ell_N$ are referred to as 'terminal nodes.' Each path along the branches which connect the root to a terminal node is labeled with one of the $q^{\ell_N}$ sequences of $q$-ary symbols of length $\ell_N$, with a one-to-one correspondence between the terminal nodes and these sequences. Assume now that a first word of length $\ell_1 < \ell_N$ is chosen as belonging to an instantaneous code. It corresponds to it a unique path from the root to some node of level $\ell_1$. Then the prefix condition forbids that sequences beyond this node be included in other codewords, meaning that the partial tree having this node as root and terminal nodes common with those of the original tree must be deleted in order to account for the choice of the first word of the instantaneous code. This entails that the terminal nodes of the partial tree, hence $q^{\ell_N - \ell_1}$ terminal nodes of the original tree, are deleted. Choosing another word of length $\ell_2$ similarly entails that $q^{\ell_N - \ell_2}$ of the remaining terminal nodes are further deleted, and this process is repeated. The number of deleted terminal nodes of the original tree cannot exceed their total number $q^{\ell_N}$, hence:

$$\sum_{i=1}^{N} q^{\ell_N - \ell_i} \leq q^{\ell_N},$$

and (5.4) results from dividing both sides by $q^{\ell_N}$.

McMillan has shown that the Kraft inequality (5.4) is actually much more general, being satisfied by *any* uniquely decipherable code regardless of the means which endow it with this property.

### 5.2.3   FUNDAMENTAL THEOREM OF SOURCE CODING

The fundamental theorem of source coding can be stated as follows. The messages generated by a stationary source, with or without memory, of entropy $H$ shannons per symbol, can be transformed according to a uniquely decipherable source encoding process into shorter (compressed) ones from which the original messages can be exactly recovered. The ratio $\lambda = \bar{\ell}/m$ of the average length after coding $\bar{\ell}$ to that before coding, $m$, i.e., the average number of symbols of the compressed message per symbol of the original one, is at least $\lambda = H/\log_2 q = H_q$, where $q$ denotes the alphabet size (assuming to be the same in both the original and the compressed messages) and $H_q$ is the entropy of the source expressed using logarithms to the base $q$. By encoding extensions of the source of increasing order $k$, $\bar{\ell}/m$ can approach as closely as desired its lower bound $H_q$.

First notice that the compression factor $\lambda$ cannot be less than $H_q$. Indeed, the largest possible entropy of a $q$-ary source is $\log_2 q$ shannons, so the source which results from the source encoding of the original source has at most this entropy. Then the average information quantity borne by a message after source coding is at most $\bar{\ell} \log_2 q$ shannons. Each message generated by the original source bears in the average an information quantity equal to $mH$ shannons. An encoding process does not create, and can at best avoid losing, information. The most efficient source coding thus results in an average length $\bar{\ell}$ such that:

$$\bar{\ell} \log_2 q \geq mH \ . \tag{5.5}$$

Dividing both sides by $m \log_2 q$ results in

$$\lambda \geq H_q \ ,$$

as stated.

This lower bound can be approached as closely as desired by a source coding process operating on a source extension of arbitrarily increasing order $k$. In order to prove this statement, we consider the source coding of the $k$-th extension of the original source. A word of a variable-length code is associated with each symbol of this extension. Let $\ell_i^{(k)}$ denote the length of the codeword which encodes the $i$-th symbol of the extension, $1 \leq i \leq q^k$, $\bar{\ell}^{(k)}$ denote its average, and $p_i^{(k)}$ denote its probability. Since $\ell_i^{(k)}$ represents a single symbol of the extension, we have $m = 1$ in (5.5). Provisionally assuming that $\bar{\ell}^{(k)}$ equals its lower bound results in

$$\bar{\ell}^{(k)} = H_q^{(k)} \ ,$$

where $H_q^{(k)}$ denotes the entropy of the $k$-th extension of the original source. This equality implies

$$\sum_{i=1}^{q^k} p_i^{(k)} \ell_i^{(k)} = -\sum_{i=1}^{q^k} p_i^{(k)} \log_q p_i^{(k)} \ .$$

The condition expressed by this equality is fulfilled only if

$$\ell_i^{(k)} = -\log_q p_i^{(k)}$$

for all possible values of $i$, but this equality does not in general result in $\ell_i$ being an integer as the length of a codeword should be. However, we may look for a set $\{\ell_i\}$ of $q^k$ integers which for the purpose of source coding approximates the set of $q^k$ positive real numbers $\{-\log_q p_i^{(k)}\}$.

It is always possible to find an integer $\ell_i^{(k)}$ such that

$$\ell_i^{(k)} \leq -\log_q p_i^{(k)} < \ell_i^{(k)} + 1$$

for each value of $i$ $(1 \leq i \leq q^k)$. Multiplying the terms of this double inequality by $p_i^{(k)}$ and summing for $i$ from 1 to $q^k$ results in

$$\sum_{i=1}^{q^k} p_i^{(k)} \ell_i^{(k)} \leq -\sum_{i=1}^{q^k} p_i^{(k)} \log_q p_i^{(k)} < \sum_{i=1}^{q^k} p_i^{(k)} \ell_i^{(k)} + \sum_{i=1}^{q^k} p_i^{(k)}$$

which can be rewritten

$$\overline{\ell}^{(k)} \leq H_q^{(k)} < \overline{\ell}^{(k)} + 1 \, ,$$

where $\overline{\ell}^{(k)}$ is the average length of the codewords; the rightmost '1' results from $\{p_i^{(k)}\}$ being a probability distribution, hence $\sum_{i=1}^{q^k} p_i^{(k)} = 1$. Dividing by $k$ results in

$$\lambda^{(k)} \leq H_q^{(k)}/k < \lambda^{(k)} + 1/k \, ,$$

where $\lambda^{(k)}$ is defined as $\lambda^{(k)} = \overline{\ell}^{(k)}/k$. If the original source is memoryless, $H_q^{(k)} = kH/\log_2 q$, where $H$ is the entropy of the original source expressed in shannons, and the double inequality above becomes

$$\lambda^{(k)} \leq H_q < \lambda^{(k)} + 1/k \, ,$$

so $\lambda^{(k)}$ approaches $H_q$ as $k$ approaches infinity. If the original source is not memoryless, remember that we defined the entropy of such a source as the limit of $H_q^{(k)}/k$ as $k$ approaches infinity (see Eq. (5.3) above), so the same conclusion holds.

The recourse to an extension of the source of high-order results in an exponential increase of the alphabet size, hence enables a finer and finer matching of the integer lengths of the codewords to the probabilities of its symbols. Most is gained, however, in the first extension orders. For instance, in the above example of Tab. 5.1, the compression factor $\lambda^{(3)} = 1.598/3 = 0.53266\ldots$ obtained with the 3-rd extension of the source is fairly close to the entropy of the original source in shannons, $0.469\ldots$, i.e., the theoretical limit.

There exist many source coding algorithms—infinitely many, in fact, since most of them depend on numerical parameters. Some of them are adaptive, i.e., do not demand that the probabilities which characterize the source operation are exactly known. These probabilities are estimated from measured frequencies, the measurement of which is an integral part of the algorithm operation. These probabilities may even be *a priori* completely unknown. Similarly, the assumption of stationarity need not be strictly satisfied, and significant compression factors can be obtained with slowly varying sources, or sources varying by steps. Given an arbitrary sequence there is no general means to know what algorithm, if any, can the most efficiently compress it.

## 5.3   ABOUT CHANNEL CODING

### 5.3.1   FUNDAMENTAL THEOREM OF CHANNEL CODING

The proof of the fundamental theorem of channel coding is rather difficult, which makes giving a convincing sketch of its proof without lengthy and somewhat technical mathematical details rather challenging.

That errorless communication cannot involve a source entropy larger than the channel capacity is rather intuitive. The mere possibility of communicating through a channel an information quantity at most equal to its capacity $C$ does not solve the problem of reliably communicating through this channel the message generated by a source of entropy smaller than or equal to this capacity. The expression of mutual information:

$$I(X; Y) = H(X) - H(X|Y) \tag{5.6}$$

appears as the difference between two terms: the information quantity $H(X)$ present at the channel input minus the uncertainty as regards $X$ which remains when $Y$ is given, measured by $H(X|Y)$. Clearly, the actual communication of a message implies that this term be zero or negligible when $H(X)$ measures the information from the source which should be received by the destination. In other words, the messages delivered to the destination should satisfy a criterion of recovery quality such that $H(X|Y)$ be zero or negligible. But $H(X|Y)$ solely depends on the channel once the probability distribution of $X$ has been given. If the channel is affected by errors, $H(X|Y)$ has some nonzero value which cannot in general be neglected. Then for faithfully communicating the messages it generates, the source may not be directly connected to the channel input. Its output should be transformed according to some coding process and the inverse transformation should be performed on the channel output so as to recover the message. Intermediate devices, namely an encoder and a decoder, should thus be interposed between the source and the channel input on the one hand, between the channel output and the destination on the other hand.

**Figure 5.3:** The channel is preceded by an encoder receiving the input variable $U$ and followed by a decoder delivering the output variable $V$. The channel input and output are $X$ and $Y$, respectively.

Figure 5.3, again an avatar of Shannon's paradigm, depicts the situation. The output of a memoryless source is connected to the input of an encoder, the output of which is connected to the input of a memoryless channel. The channel output is connected to a decoder, the output of which is the input to the destination. We assume that the same $q$-ary alphabet is used everywhere in the figure. The channel has as input and output the $q$-ary variables $X$ and $Y$, respectively. The mean mutual information $I(X; Y)$ is expressed by Eq. (5.6) with $H(X|Y) > 0$, depending on the channel. Let $U$ and $V$ denote the random variables at the input of the encoder and at the output of the decoder, respectively. Then the relation homologous to Eq. (5.6) is true:

$$I(U; V) = H(U) - H(U|V) .$$

The inequality

$$I(U; V) \leq I(X; Y)$$

is true because the encoder and the decoder do not create, but at best avoid to lose, information. The equality thus obtains for a well-designed encoding system, so

$$H(X) - H(X|Y) = H(U) - H(U|V) .$$

The criterion of recovering quality demands $H(U|V) < \varepsilon$, where $\varepsilon$ is a given positive constant smaller than $H(X|Y)$, which results in:

$$H(X) - H(U) = H(X|Y) - H(U|V) > 0 .$$

The entropy $H(U)$ must thus be less than $H(X)$, which means that the encoding is necessarily redundant.

This argument needs further comments. The random variable $X$ can actually be given two different meanings. As the channel input, it should be considered as a $q$-ary random variable since the channel deals separately with each of its input symbols. However, the source and the encoder taken together can be considered as a new source. Contrary to the original source which is assumed memoryless, the new source has memory since the encoder acts on sequences of input symbols, not separately on each of them. In the above expressions, $X$ should be understood as the *memoryless restriction* of the encoder output[2]. If on the contrary $X$ is interpreted as the output of the source with memory made of the original source and the encoder, say $X_{\text{seq}}$, then by the same token as above we have

$$H(U) \geq H(X_{\text{seq}}) ,$$

with equality for a well-designed encoder. For short, the channel input must thus be interpreted as made of independent symbols, hence represented as the $q$-ary random variable $X$, although as the output of the channel (hence at the decoder input) it consists of a sequence $X_{\text{seq}}$ having at most the same entropy as the random variable $U$ at the encoder input.

The channel capacity $C$ is the largest possible value of $I(X;Y)$. If we assume an errorless communication, $U = V$ implies that $H(U|V) = 0$. The equality of $I(X;Y)$ and $I(U;V)$ entails that

$$C \geq H(X) - H(X|Y) = H(U) ,$$

showing that errorless communication cannot be performed if the source entropy $H(U)$ exceeds the channel capacity $C$.

## 5.3.2   CODING FOR THE BINARY SYMMETRIC CHANNEL

It is more difficult to prove the positive statement of the theorem. Shannon's proof of the channel coding theorem relies on the extraordinary idea of *random coding* [72]. Since no one knew (and no one still knows) how to design the best possible code, and *a fortiori* no one could compute the probability of erroneous recovery associated with it, Shannon considered, instead of a single code, a broad set of randomly chosen codes. Then he computed the average error probability for this ensemble of

---

[2]By 'memoryless restriction,' we mean that the mutual dependence between the sequence symbols is ignored.

codes and he showed that, provided the condition of the theorem is met (i.e., the source entropy is smaller than the channel capacity), then this average error probability can be made arbitrarily small by indefinitely increasing the codeword length. The ensemble of random codes contains at least a code as good as the average, which shows that 'errorless' communication is possible. *Errorless* should be understood in an asymptotic sense, i.e., the probability of error can be made as small as desired by increasing the codeword length. The original proof given by Shannon was criticized by pure mathematicians who pointed out its lack of formal rigour and expressed doubts about the validity of the result itself. American mathematicians were especially sceptical, and the first mathematically sound proofs of the theorem came years later from the Soviet Union [50]: Shannon's conclusions were fully confirmed but the conditions of validity of the theorem were more precisely stated.

Let us first present an argument which will give us some insight, although it is not entirely satisfactory. As a simple case, we consider the binary symmetric channel already mentioned in Sec. 3.3.3. We consider sequences of a given length $n$ and we first assume that the number of errors is somehow kept constant: exactly $t$ bits, chosen at random among the $n$ bits of a sequence, are always wrong. This is a very unrealistic assumption, but it will be useful for our purpose and we shall see later that it is approximately true when $n$ and $t$ are very large. The bit error probability of the channel is thus $p_{su} = t/n$, a constant, where the subscript 'su' stands for 'substitution' since in case of an error a wrong bit is substituted for the correct one. Thanks to this simplifying assumption, the total number of different erroneous words possibly received when a given codeword is sent is exactly the number of different possible choices of $t$ objects among $n$, or combinations, generally denoted by $\binom{n}{t}$. Each of the possible transmitted $n$-bit sequences, i.e., each codeword, can thus be transformed by the channel errors into one among $\binom{n}{t}$ other $n$-bit sequences. No more than $M = 2^n / \binom{n}{t}$ distinguishable codewords can thus exist. But

$$\binom{n}{t} = \frac{n!}{t!(n-t)!}$$

and it is shown in Appendix 5.B that Stirling's formula results in the approximation:

$$\binom{n}{t} \approx 2^{n\mathcal{H}_2(t/n)} ,$$

valid for $n$ and $t$ large enough.

The maximum number of distinguishable codewords, $M = 2^n / \binom{n}{t}$, has thus the approximate expression:

$$M \approx 2^{n[1-\mathcal{H}_2(t/n)]} .$$

The entropy $H$ of a source connected to the channel input can achieve errorless communication if it is less than $(1/n)\log_2(M)$, thus approximately if:

$$H < 1 - \mathcal{H}_2(t/n) .$$

The right-hand side equals the channel capacity of the binary symmetric channel of error probability $p_{su}$ given by Eq. (3.11), so the equality

$$H < 1 - \mathcal{H}_2(p_{su}) \tag{5.7}$$

holds true asymptotically for $n$ approaching infinity.

The assumption that exactly $t$ bits are in error in any received word becomes itself asymptotically true as $t$ and $n$ approach infinity. Indeed, when the errors occur at random with probability $p_{su}$, the error frequency tends to a constant $p_{su}$ when $n$ approaches infinity as a mere consequence of the (weak) law of large numbers.

Although the calculations based on Stirling's formula are not especially intuitive, one can give a geometric interpretation of the above argument. The space of binary sequences of length $n$ contains $2^n$ elements, or points, and we may define a distance (the Hamming distance) between two of them as the number of positions in the sequence where they differ. Assuming that exactly $t$ errors occur means that there is a distance of $t$ between the transmitted sequence and the received one. Let us define the 'volume' of some subset of this space as the number of sequences, or points, it contains. The volume of the whole space is $2^n$. We may think of the set of $\binom{n}{t}$ possible received sequences when some sequence $\underline{c}$ is transmitted as an 'error sphere' of radius $t$ centered in $\underline{c}$. The volume of this sphere is $\binom{n}{t}$. If we use $M$ codewords, the total volume of the error spheres, $\binom{n}{t}M$, cannot exceed the volume of the whole space, namely, $2^n$. A geometric interpretation of inequality (5.7) results.

This argument actually shows that it is impossible to achieve errorfree communication if the source entropy is larger than the channel capacity (a conclusion we already arrived at by other means in Sec. 5.3.1). But can we show that it can be achieved if the entropy is smaller than the capacity, however, close to it may be? Intuitively, the error spheres should be as equally distributed in the whole space as possible. No explicit rule is in general available to this end. However, it turns out that choosing the codewords at random results, asymptotically as $n$ approaches infinity, in the most regular possible configuration. By 'choosing at random,' we mean that each bit of a sequence is drawn with probability 1/2 independently of the choice of other bits, and that each $n$-symbol sequence is thus chosen independently of the other sequences. The obtained regularity may look surprising since any random choice is completely unexpected hence highly irregular. Do not forget, however, that when very many random choices occur with constant probabilities, a statistical regularity appears. Probability theory states precise laws which concern random choices successively effected very many times. The 'laws of large numbers,' roughly speaking, tell that the observed frequencies tend to the probabilities when the number of considered occurrences increases, and moreover tell how closely and how fast they do.

As another simple example, the binary erasure channel considered above has the very simple capacity [Eq. (3.12)] which is quite different from that of the binary symmetric channel, [Eq. (3.11)]. Let us try to understand why. First of all, we may notice that, if we choose arbitrary binary symbols at the $n_{er}$ places where the original ones have been erased, only $n_{er}/2$ erroneous bits result in the average. This shows that, in some sense, an erasure is equivalent to half an error. The capacity of the erasure channel with erasure probability $p_{er} = 2p_{su}$ cannot thus be less than that of the binary symmetric channel of error probability $p_{su}$. It is even significantly larger since the location of the unidentified symbols is an information in itself which is destroyed when arbitrary binary symbols replace the signs $\epsilon$ which pinpoint the locations where the erasures occurred. Destroying information contained in the channel output necessarily reduces its capacity. Notice that the 'equivalence' of two erasures with a single error only holds in the binary case. If the alphabet size is $q$, the number of erroneous symbols which result in the average from arbitrary choices becomes $(q-1)n_{er}/q$ so an erasure is then 'equivalent' to $(q-1)/q$ errors.

The simplest interpretation of this case is that the number of dimensions of the space of sequences, i.e., the length $n$ of these sequences, is reduced in the average by the factor $p_{er}$. The largest information quantity that a sequence of length $n$ can bear, $n \log_2 q$ shannons, is thus reduced by the occurring erasures to $p_{er}n \log_2 q$.

### 5.3.3   GENERAL CASE: FEINSTEIN'S LEMMA

Up until now, we just gave the sketch of a proof of the fundamental theorem of channel coding for the particularly simple example of the binary symmetric channel. For any source and channel, provided they satisfy the regularity conditions of being stationary and ergodic, the fundamental theorem of channel coding states that the inequality

$$H < C \tag{5.8}$$

is the necessary and sufficient condition of asymptotically errorfree communication, where $H$ denotes the source entropy and $C$ the channel capacity.

The simple case of the binary symmetric channel is merely a particular instance of a general result named *Feinstein's lemma*. Assuming a stationary and ergodic source and a discrete stationary channel with finite memory, it states that a one-to-one correspondence between the typical sequences generated by the source (see Sec. 5.1.2) and asymptotically disjoint sets of points in the space of the output sequences can be established provided the source entropy is less than the channel capacity, when the sequence length approaches infinity. The 'error spheres' in the case of the binary symmetric channel are mere examples of such asymptotically disjoint sets of points in the channel output space. The proof of Feinstein's lemma is unfortunately rather difficult and implies lengthy mathematical developments. The statement of the fundamental theorem of channel coding then results from the lemma in a straightforward manner: observing the channel output enables uniquely identifying the channel input sequence if it belongs to the set of typical sequences. If it does not (an event of vanishingly small probability according to the Shannon-McMillan theorem), a particular distinguishable sequence can be transmitted in order to warn the destination that an atypical sequence occurred.

## 5.4   A SHORT INTRODUCTION TO THE ALGORITHMIC INFORMATION THEORY

As a convenient notation, sequences will be denoted in this section by an underlined letter to distinguish them from single symbols or numbers.

### 5.4.1   PRINCIPLE OF THE ALGORITHMIC INFORMATION THEORY

Up until now, we defined all the quantities of information theory, and especially the entropy of a source, in terms of probabilities. A very different point of view consists of measuring the information borne by a sequence by its *algorithmic complexity*, defined as the smallest length of a program which can describe this sequence. This program is written in a given alphabet, assumed here binary, and instructs a given machine to generate the sequence.

This idea was introduced by the great mathematician Kolmogorov [51], to whom we owe the axiomatization of probability theory, but also, independently, by Gregory Chaitin who was then 15 years old. Chaitin recently wrote a very readable book which popularizes the algorithmic information theory and discusses its philosophical impact [25].

Contrary to intuition, the algorithmic complexity of a sequence is almost independent of the language and of the machine used in order to describe it. Then, it becomes possible to associate with a single sequence a measure of information which does not rely on probabilities. One may think of the notion of algorithmic complexity as *prior* to that of probability and maybe able to found its theory. Let us first give a few examples of sequences and consider how they can be described.

a) The sequence

1010101010101010101010101010101010

is made of the periodic repetition of '10' 16 times. Clearly, a program describing this sequence may be as short as: "Print 16 times '10' ." An equivalent way of describing this sequence is "Print the first 32 digits of the development of 2/3 in binary numeration." Would the sequence consist of repeating '10' infinitely many times, then these descriptions become: "Print repeatedly '10' " or "Print the digits of the binary development of 2/3."

b) The sequence:

01101010000010011111001100110011

is not periodic and looks random. However, it consists of the first 32 binary digits of the development of the irrational number $\sqrt{2} - 1$, so the program "Print the first 32 digits of the binary development of $\sqrt{2} - 1$" suffices for describing it. Indefinitely continued, this sequence is described by the shorter program "Print the digits of the binary development of $\sqrt{2} - 1$."

c) The sequence:

01001010101111000011101100101100

was obtained by drawing the bits '0' and '1' at random with the same probability 1/2, independently of each others. It is likely that no program for its description exists which is much shorter than

"Print '01001010101111000011101100101100' "

and, furthermore, one could not describe a continuation of this sequence without lengthening the program, for instance by simply writing the following outcomes.

The first two sequences, (a) and b), have thus a description almost independent of their length, at variance with sequence (c) for which one may not expect to find a description much shorter than itself.

d) The sequence

01101011111111011011111100101110

also resulted from drawing '0's and '1's at random, but the probability of a '1' was 3/4 instead of 1/2. One may give a description of it shorter than its length $n$ by counting the number of '1's, say $k$, and by indicating the rank $\rho$ of this sequence in the list of all binary sequences of length $n$ comprising $k$ '1's, arranged in lexicographic order. For example, with $n = 5$ and $k = 3$, this list would be:

00111, 01011, 01101, 01110, 10011, 10101, 10110, 11001, 11010, 11100,

which comprises $\binom{5}{3} = 10$ entries. Describing the original sequence thus demands[3] a number of bits equal to $\log_2 n + \log_2 \binom{n}{k}$. For $n$ and $k$ large enough, $\binom{n}{k}$ is close to $2^{n\mathcal{H}_2(k/n)}$, where $\mathcal{H}_2(x) = -x \log_2 x - (1-x) \log_2(1-x)$ is the binary entropy function defined above by Eq. (3.6), as shown by applying Stirling formula (see Appendix 5.B). In this case, describing the sequence does not demand that the program length be equal to that of the sequence itself, $n$, but this length approximately equals $n$ times the binary entropy $\mathcal{H}_2(k/n)$, which is the smaller, its argument is more different from 1/2. One could as well use a source encoding process, for instance dealing with an extension of the sequence by the Huffman algorithm, which would result in a random equiprobable sequence, similar to sequence c), of length close to $n\mathcal{H}_2(k/n)$, too (see Sec. 5.2).

We said above that the algorithmic complexity is almost independent of the machine and of the language used in order to describe a sequence. More precisely, the machine should be universal, i.e., able to simulate the operation of any computer. Any modern general purpose computer belongs to this category. Using a given universal machine and a given language as references makes the length of a program for describing a sequence a well defined quantity. The Turing machine[4] is the prototype of such a *universal machine*. This machine can read, erase or write symbols on strips of paper. It executes a *program* written in a certain alphabet assumed here binary. This program $\underline{p} = p_1\, p_2\, p_3\, \ldots$ is written on a strip of paper (the input strip) which is always read in the same direction. At a given step of executing the program, the machine reads a bit on this strip, changes its internal state according to its own instructions, reads, or erases and writes, a bit on a second strip of paper, the working strip, which it may shift one position to the left or the right, and possibly writes on a third strip: the output strip. It finally moves one position forward on the strip which contains the program in order to read its following bit. The program is read always in the same direction, with no possible return, which results in the programs constituting an *irreducible* set in the sense that none of them is the prefix of another one (see Sec. 5.2). Indeed, when the execution of one of

---

[3]The number of bits needed to represent $k$ and $\rho$ should be understood as the smallest integer larger than or equal to the written logarithms. Representing both $k$ and $\rho$ by a fixed number of bits is necessary to unambiguously separate them; the figures given result from $k \leq n$ and $\rho \leq \binom{n}{k}$.

[4]This machine is useful in abstract reasonings but its practical use would be inconvenient and slow.

the programs ends, the machine stops and thus does not read the strip beyond the last symbol of the program just executed. As an important consequence, all the possible programs to the Turing machine have lengths which obey the Kraft-McMillan inequality (5.4).

Then the algorithmic complexity $K(\underline{x})$ of a sequence $\underline{x}$ is used as a measure of the information quantity it bears. More precisely, the average value of $K(\underline{x})/l(\underline{x})$, where $l(\underline{x})$ denotes the length of the sequence $\underline{x}$, taken in the limit as $l(\underline{x})$ approaches infinity, is an algorithmic measure of information homologous to the entropy of a source in conventional information theory but it is associated with a single sequence, not with a source, i.e., a set of sequences.

## 5.4.2   ALGORITHMIC COMPLEXITY AND ITS RELATION TO RANDOMNESS AND ENTROPY

Let $\underline{x}$ denote a finite binary sequence of length $l(\underline{x})$. Let $\mathcal{U}$ be the universal machine used as reference, e.g., the Turing machine. The sequence generated by the machine when it executes some program $\underline{p}$ is denoted by $\mathcal{U}(\underline{p})$; we assume it is a finite binary sequence.

The algorithmic complexity (also referred to as Kolmogorov complexity) of the sequence $\underline{x}$ with respect to the universal machine $\mathcal{U}$ is defined as:

$$K_{\mathcal{U}}(\underline{x}) = \min_{\underline{p}:\ \mathcal{U}(\underline{p})=\underline{x}} l(\underline{p}) . \tag{5.9}$$

In other words, it is the shortest length of a program such that the machine prints the sequence $\underline{x}$ and stops. The conditional complexity $K_{\mathcal{U}}(\underline{x}|l(\underline{x}))$ is defined the same way, but it is further assumed that the length $l(\underline{x})$ is available to the machine $\mathcal{U}$ prior to processing.

We have seen in Sec. 5.4.1 that very long sequences could be printed by a machine executing a short program such as "Print the first $n$ bits of the development of $\sqrt{2} - 1$." Translated into a binary language, this program is much shorter than $n$ if $n$ is large, and its length weakly depends on $n$. On the contrary, describing a random sequence demands a program approximately as long as the sequence itself. Here are a few important results as regards measuring the algorithmic complexity of a sequence by a program length.

If $\mathcal{U}$ is a universal machine and if $\mathcal{A}$ is another machine, then

$$K_{\mathcal{U}}(\underline{x}) \leq K_{\mathcal{A}}(\underline{x}) + c_{\mathcal{A}} , \tag{5.10}$$

where $c_{\mathcal{A}}$ is a constant independent of $\underline{x}$. This constant is the length of the program which teaches the universal computer $\mathcal{U}$ how to simulate $\mathcal{A}$. If $\mathcal{A}$ is also a universal machine, then $|K_{\mathcal{U}}(\underline{x}) - K_{\mathcal{A}}(\underline{x})| < c$, where $c$ is a constant independent of $\underline{x}$. This property establishes that the complexity is *universal*, in the sense that the length difference between the programs describing $\underline{x}$ with machines $\mathcal{U}$ and $\mathcal{A}$ is negligible if $\underline{x}$ is large enough. The subscript which tells what machine has been used in order to define the complexity may thus be dropped, as it will be in the sequel.

The conditional complexity of $\underline{x}$ is smaller than its length, except for a constant:

$$K(\underline{x}|l(\underline{x})) \leq l(\underline{x}) + c . \tag{5.11}$$

This result is obvious since the program may contain the sequence $\underline{x}$ itself. If the length of $\underline{x}$ is not known, hence for nonconditional complexity, one has the rough upper bound:

$$K(\underline{x}) \leq l(\underline{x}) + 2 \log_2 l(\underline{x}) + c . \tag{5.12}$$

In order to prove it, it suffices to return to the previous case by feeding the machine with the length $l(\underline{x})$, besides $p$. One needs $\log_2 l(\underline{x})$ bits for expressing this length, but the number of bits which is necessary for this purpose is itself unknown. If one repeats each bit twice consecutively [which accounts for the factor 2 in (5.12)], it becomes possible to unambiguously mark the end of the binary representation of this number with the doublet '01.' More subtle means can be used for still shortening the expression of $l(\underline{x})$, resulting in a slightly smaller bound.

Another important result is that the low complexity sequences are few, in the sense that:

$$\#\{(\underline{x} \in (0, 1)^* : K(\underline{x}) < k)\} < 2^k \tag{5.13}$$

where $\#\{\cdot\}$ denotes the number of elements of a set and where $(0, 1)^*$ represents the set of possible binary sequences of any length (i.e., $\lambda$, 0, 1, 00, 01, 10, 11, 000, $\ldots$, where $\lambda$ denotes the empty sequence). In other words: the total number of sequences of algorithmic complexity less than $k$ is less than $2^k$. The proof of this result is obtained by merely counting the programs of length $k$. Their number is less than $2^k$, and each of them can produce at most a sequence $\underline{x}$.

Still another result on algorithmic complexity is the bound:

$$K(\underline{x}|n) \leq n\mathcal{H}_2 \left( \frac{1}{n} \sum_{i=1}^{n} x_i \right) + 2 \log_2 n + c , \tag{5.14}$$

where $\underline{x} = x_1 x_2 \ldots x_n$ is a binary sequence of length $n$. $\mathcal{H}_2(\cdot)$ is the binary entropy function defined by Eq. (3.6). The symbols of $\underline{x}$ are dealt with as integers in (5.14), so $\sum_{i=1}^{n} x_i$ involves the ordinary addition and merely equals the number of '1's in $\underline{x}$ (its weight). This bound may be proved using a construction similar to the example of compressing the binary sequence drawn with a probability of bit '1' different from 1/2, given in Sec. 5.2.

The algorithmic complexity of a random sequence (per symbol) is close to its entropy. In order to prove this statement, one first notices that the set of programs have lengths which satisfy the Kraft-McMillan inequality, i.e.,

$$\sum_{p: \, \mathcal{U}(\underline{p}) \text{ stops}} 2^{-l(\underline{p})} \leq 1 . \tag{5.15}$$

Let $\underline{X} = X_1 X_2 \ldots X_n$ be a sequence of length $n$ of random variables, each drawn at random independently of the others with a same probability distribution $f(x)$, where $x$ is an element of some finite alphabet of size $q$. Sequences of length $n$ have as probability distribution $\prod_{i=1}^{n} f(x_i)$, which is denoted by $f(x^n)$. Then there exists a constant $c$ such that

$$H(X) \leq \frac{1}{n} \sum_{x^n} f(x^n) K(\underline{X}|n) \leq H(X) + \frac{q \log_2 n}{n} + \frac{c}{n} \tag{5.16}$$

for any $n$, where $\sum_{x^n}$ means that the sum is effected for all sequences of length $n$. $H(X)$ denotes the entropy associated with the distribution $f(x)$. The expectation of the complexity $K(\underline{X}|n)$, namely $\sum_{x^n} f(x^n)K(\underline{X}|n)$, tends to the entropy $H(X)$ as $n$ tends to infinity since then the two rightmost terms vanish.

Difficulties nevertheless exist so one may not liken without precautions Shannon's entropy and the algorithmic complexity per symbol. For example, the familiar formulas

$$H(X, Y) = H(Y, X)$$

and

$$I(X; Y) = H(X) - H(X|Y)$$

have an equivalent for algorithmic complexity except for an additional term which is of the same order of magnitude as the logarithm of their left-hand sides [51].

Let $\underline{X} = X_1 X_2 \ldots X_n$ be a sequence of binary independent random variables of probability 1/2. One then has:

$$\Pr[K(\underline{X}|n) < n - k] < 2^{-k} , \tag{5.17}$$

for any $k < n$. This result means that most sequences have a complexity which is close to their length.

A sequence $\underline{x} = x_1 x_2 \ldots x_n$ is said to be *algorithmically random* if

$$K(\underline{x}|n) \geq n . \tag{5.18}$$

Counting the sequences shows that there exists at least one such sequence for each value of $n$.

A sequence $\underline{x} = x_1 x_2 \ldots x_n$ is said to be *incompressible* if

$$\lim_{n \to \infty} \frac{K(\underline{x}|n)}{n} = 1 . \tag{5.19}$$

The law of large numbers implies for incompressible sequences:

$$\frac{1}{n} \sum_{i=1}^{n} x_i \to \frac{1}{2} , \tag{5.20}$$

meaning that the proportion of the '0's and the '1's in such a sequence is almost the same.

For a random binary sequence $\underline{X}$ of length $n$ with unequal probabilities of '0' and '1,' $\Pr(X = 1) = \theta \neq 1/2, 0 < \theta < 1$, one obtains that

$$\Pr[\frac{1}{n} K(\underline{X}|n) - \mathcal{H}_2(\theta) \geq \varepsilon] \to 0, \ n \to \infty . \tag{5.21}$$

### 5.4.3   SEQUENCES GENERATED BY RANDOM PROGRAMS

Let us now consider a universal machine fed by a *random* program. By random program, we mean that each of its bits is chosen at random with probability 1/2, independently of the other bits. There exists a nonzero probability that such a randomly chosen program be significant for the machine and thus eventually results in the machine printing some output sequence.

But how does the execution of a program stop? A theorem of theoretical computer science (quite similar to Gödel's theorem) states that no algorithm can predict whether the execution of a program will stop after printing a sequence, or will continue to run indefinitely without providing an output. This theorem implies that the algorithmic complexity of a sequence $x$ is *uncomputable*. The only means to find the shortest possible program describing a sequence $x$ would be to try all possible programs, but the execution of certain of them would never stop.

From a point of view closer to that of conventional information theory, the uncomputability of a sequence complexity of is not so surprising. We actually noticed in Sec. 5.2 that infinitely many possible source coding algorithms exist. It is thus impossible, when the description $p$ of a sequence $x$ has been obtained using some source coding algorithm, to assert that a shorter one cannot be obtained using another algorithm, for lack of being able to try all algorithms. One can only assert that $K(x) \leq l(p)$.

Chaitin introduces a number he denotes by $\Omega$ [25], equal to the probability that the execution of a program stops, namely:

$$\Omega = \sum_{p:\, \mathcal{U}(p) \text{ stops}} 2^{-l(p)} . \tag{5.22}$$

Since the lengths $l(p)$ of possible programs satisfy the Kraft-McMillan inequality (5.4), one actually defines by this means a number between 0 and 1. There exists no effective means for calculating $\Omega$, just like the complexity of a sequence cannot be computed.

The probability of a random binary program $p$ of length $l(p)$ is $\Pr(p) = 2^{-l(p)}$, assuming its bits are chosen independently of each other with probability 1/2. A short program thus more probably occurs than a long one. If a short program produces a long sequence, this sequence cannot be algorithmically random since it has a simple description. One is led to define the *universal probability* of a sequence $x$ as:

$$P_{\mathcal{U}}(x) = \sum_{p:\, \mathcal{U}(p) = x} 2^{-l(p)} = \Pr[\mathcal{U}(p) = x] . \tag{5.23}$$

In this case, too, this equality defines a probability since the lengths $l(p)$ of the possible programs satisfy the Kraft-McMillan inequality. It is the probability that a program resulting from the choice of random successive independent bits with the same probability 1/2 results in the universal machine $\mathcal{U}$ printing the sequence $x$. This probability is not exactly independent of the machine, but only weakly depends on it.

To conclude this section, the algorithmic information theory has a great theoretical interest. The uncomputability of the algorithmic complexity makes it, however, poorly fitted to practical applications and somewhat frustrating. Moreover, the algorithmic information theory essentially appears as a theory of sequences becoming *infinitely long*. The equivalence between machines, which justifies the universality of the algorithmic complexity, is established by Eq. (5.10) only apart from adding a constant, and this constant may be of the same order of magnitude as the length of practically interesting sequences.

There is apparently a wide difference between information theory in the sense of Shannon, which is based on probabilities, and the algorithmic information theory which apparently involves a single sequence. It may thus look surprising that the algorithmic complexity and the entropy of a random sequence are close to each other, as stated in Sec. 5.4.2. Apparently, the algorithmic complexity is based on a single sequence and the entropy on a set of sequences endowed with probabilities. However, estimating probabilities involves frequency measurements, which for long sequences can only concern a very tiny fraction of all possible sequences of that length. On the other hand, the algorithmic complexity of a sequence implicitly refers to a collection of sequences, i.e., the subset of all sequences which share the same complexity.

## 5.5    INFORMATION AND ITS RELATIONSHIP TO SEMANTICS

Shannon and the pioneers of information theory had an empirical approach since they did not attempt *defining* information, nor explicating its connection with semantics. They just proposed means for its quantitative measurement. Communication engineering being concerned only with *literal communication*, it was possible to discard semantics as irrelevant. In the very first page of the paper which founds information theory [72], Shannon wrote:

> The fundamental problem of communication is that of reproducing at one point either exactly or approximately a message selected at another point. Frequently the messages have *meaning*; that is they refer to or are correlated according to some system with certain physical or conceptual entities. These semantic aspects of communication are irrelevant to the engineering problem. The significant aspect is that the actual message is one *selected from a set* of possible messages. The system must be designed to operate for each possible selection, not just the one which will actually be chosen since this is unknown at the time of design. (Shannon's italics.)

In short, just like a messenger has not to know about the message he/her carries, a communication system should ignore meaning and operate regardless of the message. Replacing in the above quotation the word 'point' which refers to a location in space with the word 'instant' moreover extends its relevance to communication in time, hence to heredity.

We try here rather naively to define information, hopefully shedding some light on the relationship between information and semantics. Contrary to the remainder of this chapter, the remarks below do not express any consensus among information theorists.

Let us consider a digital message, i.e., a sequence of symbols from some finite-size alphabet. Such a message is a mathematical abstraction which needs a physical support for having any interaction with the real world, and especially for being communicated. The physical support of a message can take a variety of forms, either material (e.g., ink deposits on a sheet of paper according to conventional patterns, i.e., letters, holes punched in a card, local states of magnetization of some ferromagnetic medium, shallow tiny holes in a compact disk (CD), etc.), or consisting of the variation of some physical quantity as a function of time (e.g., air pressure, or electrical current, or electromagnetic field, etc.). The material supports are actually used for recording, i.e., communication through time consisting of writing a message to be read later, while the variation in time of a physical quantity can be propagated as a wave hence enables communication through space. Regardless of its support, each of the alphabet symbols needs only be unambiguously distinguishable from the other ones.

A same message can be supported by different physical media which, moreover, can be converted from one to another. For instance, the message recorded on a computer memory or a CD can be converted into an acoustic wave (i.e., a sound), or emitted as an electromagnetic wave. Similarly, the alphabet size can be changed: a musical record in a computer memory uses the binary alphabet, but a large-size alphabet is needed for converting it into an audible acoustical signal. The message itself can moreover be changed so as to improve its characteristics as regards some desired property. For instance, it may be compressed in order to reduce the size of the memory needed for its recording (this is source coding as briefly considered in Sec. 5.2), or on the contrary, expanded by a redundant encoding making it resist transmission errors. This is channel coding, defined in Sec. 5.3 and a major topic of this lecture, to be developed in the next chapter.

A message can thus exist in a variety of equivalent forms, depending on its possible encoding, alphabet size, and physical support. We refer to the underlying entity which is common to all these forms as *an information*. We may thus define an information as the equivalence class of all the messages which can be converted into each other by changing their encoding, alphabet size, or physical support. Each of these messages will be said to *bear* this information. The equivalence class associated with an information clearly contains infinitely many elements. Any of these messages is a representative of this class. An information is thus an abstract object which manifests itself in the physical world by any of its representatives, or *realizations*.

As an example, the sequence of Latin letters and spaces:

Information theory discards semantics (1)

and the binary sequence

1001001 1101110 1100110 1101111 1110010 1101101 1100001 1110100 1101001 1101111
1101110 0100000 1110100 1101000 1100101 1101111 1110010 1111001 0100000 1100100
1101001 1110011 1100011 1100001 1110010 1100100 1110011 0100000 1110011 1100101
1101101 1100001 1101110 1110100 1101001 1100011 1110011

share the same *information*, since the latter just resulted from transforming sequence (1) using the ASCII (American Standard Code for Information Interchange) 'code' which is currently used in computer memories. Each Latin letter or space is replaced by a 7-bit word according to a one-to-one correspondence.

The binary sequence

10010011 11011101 11001100 11011110 11100100 11011011 11000011 11101000 11010010
11011110 11011101 01000001 11101000 11010001 11001010 11011110 11100100 11110011

01000001 11001001 11010010 11100111 11000110 11000011 11100100 11001001 11100111
01000001 11100111 11001010 11011011 11000011 11011101 11101000 11010010 11000110
11100111

also bears the same information as sequence (1) since an 8-th bit has been appended to each of the 7-bit words of the first binary sequence, equal to the sum modulo 2 of its bits thus making the total number of '1's even. This may be thought of as a rudimentary means of error control: if an error affects a symbol in a 8-bit word, the number of '1's becomes odd so counting the '1's in each word enables detecting a single-symbol error. Of course, the first binary sequence could be transformed by sophisticated error-correcting codes into an equivalent one made resilient to errors (up to a limit) and bearing again the same information.

As a counterexample, the sentence:

La théorie de l'information exclut la sémantique (2)

is a French translation of the English sentence (1). Although it looks close to it, articles and a preposition have been appended to comply with the French grammar and the word order is different, so sentence (2) does not bear the same information as (1). However, both sentences share the same *meaning*.

The multiplicity of messages which bear the same information leads to ask the following question: given some alphabet with an arbitrary physical support, is there a minimal-length realization of a given information? The simplest alphabet, i.e., the binary one, is a natural choice. The problem becomes whether a minimal-length binary message exists within the equivalence class associated with the given information. Shannon's information theory asserts this existence when the given representative is a message generated by a stationary probabilistic source. The algorithmic information theory extends this statement to any message that a universal computer can generate. We'll refer to the minimal binary realization of an information as its *information message*.

The fundamental theorem of source coding of Shannon's information theory states that any $n$-symbol message generated by this source can be transformed by source coding into a binary message of average length at least $\overline{\ell} = nH$ bits where $H$ denotes the source entropy per symbol, expressed using binary information units, i.e., shannons. This minimal-length realization of an information generated by a probabilistic source is its information message. It results from optimal source coding, which entails that its bits are probabilistically independent and equiprobable. In the algorithmic information theory, the length of the information message is used for defining the algorithmic complexity associated with the given information. The first case is much more restrictive than the second one, which may be considered as general. However, the source stationarity enables effectively estimating the probabilities of the sequences it generates, hence its entropy, by making frequency measurements. In the second case, on the contrary, the existence of a minimal-length realization is mathematically proven, but the algorithmic complexity is actually an uncomputable quantity which thus can generally not be evaluated.

The set of all binary messages of length $\ell$ can be represented by a tree like that of Fig. 5.2 with each of its branches labeled with a bit according to some convention. Let us interpret the $i$-th bit of the information message as the answer to a dichotomic question (i.e., answerable by yes-or-no), '0' meaning for instance 'yes' and '1' meaning 'no.' Any information message of length $\ell$ may thus be interpreted as one of the $2^\ell$ paths of a binary tree of length $\ell$ taken from the root to the leaves, the choice of a branch at any fork being made at random with probability 1/2. A path in this tree, i.e., an information message of length $\ell$, can be interpreted as *an integer* $i$, $0 \leq i \leq 2^\ell - 1$. The questions associated with the successive bits of the information message may for instance be those needed for identifying the species to which some given living being belongs. Provided the set of species is ordered according to a binary hierarchical taxonomy, using $\ell = nH$ properly chosen successive dichotomic questions enables distinguishing from each other $2^{nH} = (2^H)^n$ species. The entropy $H$ of the source then measures the ability of the messages it generates to discriminate among objects of some outer world, hence to bear some kind of semantics. The corresponding information quantity is simply the number of binary digits which are needed to represent it.

Establishing a correspondence between each bit of the information message and a dichotomic question makes it eventually identify or represent an object which belongs to some outer world, like living beings as in the above example of taxonomy, making possible to distinguish between them. Then information in the above meaning is given a semantic content according to an external convention which we may refer to as its *meaning*.

This is a rather crude kind of semantics, apparently restricted to representing material objects which can be ordered in a tree-like fashion. However, the possible semantic content can be widely extended if we notice that:

— Describing an outer reality by a binary message is not limited to answering dichotomic questions. Data of various other kind can also be represented. For instance, grouping $k$ bits of the information message into a block may be used to specify that the value of some parameter is one of $2^k$ predetermined levels. If $k$ is large enough, this amounts to approximately specify the value of a continuously varying parameter. This is for instance currently used in telephony to represent instantaneous values of the speech signal (referred to as samples) by a sequence of bits, a process referred to as 'pulse code modulation' (PCM). For frequent enough samples (8 kilohertz, i.e., a sample every 125 microseconds) and $k$ as low as 8 (hence $2^8 = 256$ levels), a sufficient quality of speech transmission is achieved.

— Moreover, a relation between material objects can be represented by the same means as the objects themselves, then opening the semantic field to abstract as well as material objects.

Information theory, either Shannon's or algorithmic, uses the length of the information message as a quantitative measure of information. Since an information message of length $\ell$ enables distinguishing $2^\ell$ different objects, $\ell$ is a logarithmic measure of the discriminating ability of the information message, regardless of the distinguished objects, a matter of semantics. We may thus understand a quantity of information as the number of semantic instances it can represent, or as the number of dimensions of some space which represents semantic objects. It should be kept in mind that the information quantity is by no means an exhaustive description of an information, just like the mass is only one of the many possible attributes of a material object (shape, temperature, electric charge, color, texture, internal organization, etc.).

## 5.6   APPENDICES

### APPENDIX 5.A. DETAILED ANALYSIS OF A MARKOVIAN SOURCE

We now consider the example of a binary Markovian source, as depicted in Fig. 5.1-**b**, having memory $m = 3$ hence $s = 2^3 = 8$ states, for a more detailed analysis. Let us assume that the initial content of the memory is 000. Let $x$ (0 or 1) denote the next output symbol. It changes the memory content into $00x$. The next symbol, say $y$, shifts again the content which becomes $0xy$, and one more symbol, say $z$, results in the content $xyz$ from which all the initial bits were removed. It is convenient to define the state of the memory as the integer $s = 4x + 2y + z$, which assumes values from 0 to 7 and uniquely labels the 8 possible contents of the memory. The probability of transmitting 1 in state $s$ is denoted by $p_s$, that of transmitting 0 in the same state is $q_s = 1 - p_s$. The following table plots the transition probabilities between the states.

Table 5.2: Each entry in this table is the probability of the transition from the corresponding state in the leftmost column into that of the top row

| $s$ | 0 | 1 | 2 | 3 | 4 | 5 | 6 | 7 |
|-----|-----|-----|-----|-----|-----|-----|-----|-----|
| 0 | $q_0$ | $p_0$ | | | | | | |
| 1 | | | $q_1$ | $p_1$ | | | | |
| 2 | | | | | $q_2$ | $p_2$ | | |
| 3 | | | | | | | $q_3$ | $p_3$ |
| 4 | $q_4$ | $p_4$ | | | | | | |
| 5 | | | $q_5$ | $p_5$ | | | | |
| 6 | | | | | $q_6$ | $p_6$ | | |
| 7 | | | | | | | $q_7$ | $p_7$ |

The steady probability of state $s$ is denoted by $x_s$. Thus, we have $\sum_s x_s = 1$, where the sum is extended to all values of $s$. The transitions between the states imply that:

$$x_0 = q_0 x_0 + q_4 x_4, \quad x_1 = p_0 x_0 + p_4 x_4,$$

$$x_2 = q_1 x_1 + q_5 x_5, \quad x_3 = p_1 x_1 + p_5 x_5,$$

$$x_4 = q_2 x_2 + q_6 x_6, \quad x_5 = p_2 x_2 + p_6 x_6,$$

$$x_6 = q_3 x_3 + q_7 x_7, \quad x_7 = p_3 x_3 + p_7 x_7.$$

This is a homogeneous system of linear equations, the solution of which is defined only up to a common factor. The probabilities $x_s$ which solve this system can thus all be expressed in terms of one of them, say $x_1$, and the common factor is eventually determined by solving $\sum_s x_s = 1$. We define the coefficients $\{a_s, s = 0, 1, \ldots, 7\}$ as $a_s = x_s/x_1$. Then, $a_0 = q_4/p_0$, $a_1 = 1$, $a_4 = 1$, $a_3 = a_6$, $a_7 = a_3 p_3/q_7$. The three remaining unknowns $a_2$, $a_3$ and $a_5$ satisfy the three equations:

$$a_2 + a_3 = 1 + a_5 \,,$$

$$a_2 = q_1 + a_5 q_5$$

and

$$a_5 = a_2 p_2 + a_3 p_6 \,.$$

This system has as solutions:

$$a_5 = \frac{p_1 p_6 + q_1 p_2}{1 - p_2 q_5 - p_5 p_6} \,,$$

$$a_2 = q_1 + q_5 \frac{p_1 p_6 + q_1 p_2}{1 - p_2 q_5 - p_5 p_6} = q_1 + q_5 a_5 \,,$$

and

$$a_3 = p_1 + p_5 \frac{p_1 p_6 + q_1 p_2}{1 - p_2 q_5 - p_5 p_6} = p_1 + p_5 a_5 \,.$$

The equality $\sum_s x_s = 1$ results in

$$x_1 = 1/(a_0 + 2 + a_2 + 2 a_3 + a_5 + a_7) \,.$$

All the other probabilities $x_s$ follow from $x_s = a_s x_1$. If we let $p_s = 1/2$ for any $s$ all the coefficients $a_0, a_1, \ldots, a_7$ become 1, so that all the probabilities $x_s$ become 1/8, as expected. Then the source is actually a memoryless source which transmits 0 or 1 with probability 1/2.

*Calculator simulation.* For a numerical application, we assumed that the probability of transmitting symbol '1' when the memory is in state $s$ is $p_s = [1 + 2w(s)]/8$, which only depends on the weight $w(s)$ of the state, i.e., on the number of '1's in its binary representation, and is an increasing function of it. This choice was intended to favor the clustering of either '0's or '1's in long segments. The obtained sequences are very different from that resulting from the independent choice of '0' or '1' with probability 1/2, which evidences the memory effect. Computing the state probabilities in this case results in $x_0 = x_7 = 5/16$, $x_1 = x_2 = x_3 = x_4 = x_5 = x_6 = 1/16$. One easily checks that the steady probabilities of '0' and '1' in the generated sequence are both 1/2. The operation of this Markovian source has been simulated using a calculator. Starting from the initial state $s = 0$, the following sample sequence has been obtained:

```
00000000000101000000000000000000010000000100100110111100000000000100000000011111
11000000000000000100000000000000010000000010000000000011000000100000000000101001
01011111110100010101110100111111111111101000010010001110011011011110111101111100
11110110111111111010111111111101111111111000001000000000000011011111100001000
001110110000000000000000000011110001001111111111111111111 ...
```

## APPENDIX 5.B. APPROXIMATIONS BASED ON THE STIRLING FORMULA

Factorial $n$, denoted by $n!$, is the product $1 \times 2 \times \dots \times n$ of the first $n$ positive integers. It is the number of possible different permutations of $n$ objects. An approximate expression of $n!$ valid when $n$ is large is provided by Stirling's formula:

$$n! \approx (n/e)^n \sqrt{2\pi n}(1 + \varepsilon_n) \qquad (5.24)$$

where $\varepsilon_n$ approaches $1/12n$ as $n$ approaches infinity, $e = 1 + \sum_{i=1}^{\infty} 1/i!$ is the base to the 'natural' logarithms, denoted by $\ln(\cdot)$, i.e., $\ln(e) = 1$, and $\pi$ is the ratio of the perimeter of a circle to its diameter. This approximation is asymptotically tight for $n$ large, meaning that it is comparatively the closer to the exact value, the larger $n$.

The number of combinations of $t$ objects among $n$, denoted by $\binom{n}{t}$, is equal to

$$\binom{n}{t} = \frac{n!}{t!(n-t)!} \, .$$

If we assume that both $n$ and $t$ are large, an approximate, but asymptotically tight, expression of $\binom{n}{t}$ results from Stirling formula, namely:

$$\binom{n}{t} \approx \frac{1}{\sqrt{2\pi t (n-t)}} \left(\frac{n}{t}\right)^t \left(\frac{n}{n-t}\right)^{n-t}$$

which can be rewritten as

$$\binom{n}{t} \approx 2^{n\mathcal{H}_2(t/n)-(1/2)\log_2(2\pi t(n-t))} ,$$

where $\mathcal{H}_2(\cdot)$ has been defined by Eq. (3.6). If $n$ is large enough it approaches

$$\binom{n}{t} \approx 2^{n\mathcal{H}_2(t/n)} \qquad (5.25)$$

since $\log_2(x)/x$ approaches $0$ as $x$ approaches infinity.

CHAPTER 6

# An Outline of Error-Correcting Codes

## 6.1 INTRODUCTION

This chapter provides an introduction to error-correcting codes, considered as tools for making *reliable communication through unreliable channels*. A lot of works were devoted to error-correcting codes but only a few of them are actually accounted for below. Even when dealing with well-known codes, the approach for presenting them is rather unconventional: much of the vocabulary and notation used in this text departs from the standard ones, which have their origin in the history of the discipline rather than in a logical development. Our starting point will be an analysis of mere repetition, which is a rather crude error-correcting means. Then more elaborated error-correcting codes will be introduced. Interestingly, this approach will enable us to meet and understand practically important families of error-correcting codes, including the most efficient ones. We need first to describe the basic communication problem and the kind of channels which will be considered.

## 6.2 COMMUNICATING A MESSAGE THROUGH A CHANNEL

### 6.2.1 DEFINING A MESSAGE

A *message* consists of an arbitrary sequence of elements, referred to as *symbols*, taken from some alphabet, i.e., some finite collection of signs. For instance, the previous sentence is made of letters from the Latin alphabet, plus the interval which enables separating the words. If we ignore the punctuation marks and typographic differences (some letters are in italic or capitalized), we may thus think of the alphabet as containing 27 symbols (we shall say that its size is 27). Similarly, the genomic message is made of a sequence of nucleic bases which belong to the 4-symbol alphabet {**A, T, G, C**}. The only requirement on the alphabet is that it be made of unambiguously distinguishable symbols, so the smallest possible size of an alphabet is 2. We shall assume in the following that we consider messages from the binary alphabet and we shall represent its symbols by the digits 0 and 1, often referred to as *bits*, an acronym for 'binary digits'[1]. If a given message uses an alphabet of size larger than 2, it can be converted into the binary one. For instance, if we represent the nucleic bases according to the correspondence rule **A** → 00, **G** → 01, **T** → 10 and **C** → 11, we can replace, e.g., the 6-symbol message from the nucleic-base alphabet **AATCGG** by the 12-bit binary message 000010110101, where each block of two successive bits represents a nucleic base according

---

[1] Remember that we use 'bit' with the meaning of binary symbol, never to name the binary information unit which is referred to as 'shannon.'

to the following (arbitrary) rule: the first bit represents the chemical structure of the nucleotide, 0 meaning 'purine' (**R**) and 1 'pyrimidine' (**Y**); the second bit indicates to what complementary pair the nucleotide belongs, 0 meaning **A–T** (or **W**) and 1 **C–G** (or **S**). The conversion is especially easy here since 4 is a power of 2, but conversion from an alphabet of any size into the binary alphabet is possible by slightly less simple means. If, for instance, the alphabet is of size 3, say {0, 1, 2}, the correspondence rule may be $0 \to 00$, $1 \to 01$ and $2 \to 1$. Then, a symbol 0 or 1 of the ternary alphabet is represented by a block of two bits which begins with the bit 0, while the ternary symbol 2 is represented by the single bit 1 (this rule obeys the prefix condition stated above in Sec. 5.2 hence enables unambiguously separating the binary representations of ternary symbols). Assuming the messages to be binary thus results in little loss of generality. We shall do so in the sequel for the sake of simplicity.

When a $k$-symbol binary message is represented by an $n$-symbol codeword, the average information quantity which is borne by a transmitted symbol is $R = k/n$, which is referred to as the code rate. Thus, $R$ equals the entropy of the 'redundant source' made of the channel encoder fed by the original source (see Fig. 3.8 in Sec. 3.6). The fundamental theorem of channel coding thus implies that errorless communication is possible if and only if $R = k/n < C$, where $C$ is the channel capacity in shannons. If the message symbols belong to the $q$-ary alphabet, then the above inequality becomes $(k/n) \log_2(q) < C$ since each of the $k$ information symbol now bears an information quantity of $\log_2(q)$ shannons.

## 6.2.2   DESCRIBING A CHANNEL

A *channel* is any medium where symbols can be propagated or recorded. For instance, if the symbols 0 and 1 of the binary alphabet are represented by the polarity of a voltage, say $+v$ corresponding to 0 and $-v$ to 1, measuring the polarity at the extremity of a pair of conducting wires will let know what binary symbol was represented by the voltage at the other extremity. Changing the polarity at regularly spaced intervals according to the sequence of bits of some message, at the transmitting end, will enable to reconstruct the transmitted message at the receiving end. We are interested in the case where the channel is *unreliable*, i.e., where observing the voltage at the receiving end does not ensure with certainty that the transmitted bit is identified, due to the addition to the proper voltage of a spurious one, referred to as *noise*. Then the channel is characterized by its *signal-to-noise ratio* which is the ratio of the signal power to that of the noise. For a given constant signal-to-noise ratio, the decisions taken about the binary symbols at the receiving end incur a constant probability of error which is the smaller, the larger the signal-to-noise ratio. We then obtain the simple model to be introduced now.

The simplest model of a binary channel is referred to as the *binary symmetric channel*. A channel can be described by the probabilities of the transitions between its input and output symbols, as depicted in Fig. 6.1 (already drawn as Fig. 3.4-**a** in Ch. 3). This channel is *memoryless*, i.e., deals separately with the successive input symbols. Its input symbol $X$ and its output symbol $Y$ are both binary. If $X = 0$, there is a probability $p$ that $Y = 1$, and if $X = 1$ that $Y = 0$ with the *same*[2] probability $p$, referred to as the channel error probability.

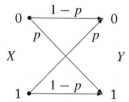

**Figure 6.1:** Binary symmetric channel with error probability $p$. All the transition probabilities from an input symbol to an output symbol are plotted on the figure.

Another description can be given of the same channel. Let $E$ denote a binary random variable which assumes the value 1 with probability $p < 1/2$ and hence the value 0 with probability $1 - p$. The assumption that $p < 1/2$ does not incur any loss of generality, since exchanging the arbitrary signs 0 and 1 suffices to replace the initial error probability $p > 1/2$ with $1 - p < 1/2$; the case $p = 1/2$ may be discarded as equivalent to the absence of any channel, since then the output does not depend on the input. Then the channel operation can be described as the addition modulo 2 of $E$ to the input binary random variable $X$, namely:

$$Y = X \oplus E \tag{6.1}$$

where $\oplus$ denotes the addition modulo 2: $0 \oplus 0 = 0, 0 \oplus 1 = 1 \oplus 0 = 1$ as in ordinary addition, but $1 \oplus 1 = 0$. (The addition modulo 2 may be interpreted as representing the parity of the ordinary addition, the binary variables '0' and '1' meaning 'even' and 'odd,' respectively.) Then, adding 1 modulo 2 results in changing a bit to its binary complement as does an error in the binary symmetric channel.

When endowed with the addition modulo 2 and the ordinary multiplication (such that $0 \times x = 0$ and $1 \times x = x$, where $x$ is any binary symbol), the binary set $\{0, 1\}$ acquires the mathematical structure of a *field*: any of its elements has an inverse for the addition, there exists a neutral element for the addition (0), any nonzero element has a multiplicative inverse. Here, 1 is the single nonzero element and is its own inverse, so it is the neutral element for multiplication. The binary set $\{0, 1\}$ is the smallest set which can be endowed with the field structure.

The error probability $p$ depends on the noise sample which affects the transmitted voltage which, for instance, has some known probability distribution (typically, a Gaussian one). At the receiving end the noise sample is not known but each received voltage is measured and used to assess the reliability of the corresponding symbol. Then the output of the channel is a real quantity, say $a$, which for Gaussianly distributed noise is related to the transmitted binary symbol $x$ and to the

---

[2]It is why the channel is referred to as 'symmetric.'

instantaneous error probability $p$ (up to a constant factor) according to:

$$a = (-1)^x \ln \frac{1-p}{p} , \tag{6.2}$$

which implies

$$p(a) = \frac{1}{\exp[(-1)^x a] + 1} = \frac{1}{\exp(|a|) + 1} . \tag{6.3}$$

By 'instantaneous error probability,' we mean the probability that an error affects the received symbol when the measured channel output is the real number $a$. Its dependence on $a$ has been explicated by the notation $p(a)$. Assuming $p(a)$ to be less than 1/2, $\ln \frac{1-p(a)}{p(a)}$ is positive so the sign of $a$ is positive for $x = 0$ and negative for $x = 1$. The sign of $a$ thus bears a binary decision. The magnitude $|a| = \ln \frac{1-p(a)}{p(a)}$ measures the reliability of this decision since it is a decreasing function of the error probability, with $|a| = 0$ for $p = 1/2$ (the decision is indeed not reliable at all in this case) and $|a|$ approaching infinity if the error probability approaches 0 (then the decision is fully reliable). It will be convenient to refer to a quantity like (6.2) as the *real value*[3] of the corresponding received binary symbol and its use is referred to as a 'soft decision.'

If the measured received voltage is not used in the subsequent process, but only the binary decision based on its sign, $p(a)$ must be replaced by its average $\bar{p}$ and the channel is described by the scheme of Fig. 6.1 with constant $\bar{p}$. The performance is significantly degraded but the process is much simplified. This case is referred to as 'hard decision.'

Reliably communicating through an unreliable channel means the ability to communicate a message (for instance, a sequence of binary symbols) with an arbitrarily small probability of error through a noisy channel (for instance, a binary symmetric channel as just described, having a symbol error probability $\bar{p} > 0$, or the channel with binary input and real output, namely, the real value $a$ of each received symbol).

## 6.3   REPETITION AS A MEANS OF ERROR CORRECTION

### 6.3.1   ERROR PATTERNS ON REPEATED SYMBOLS AND THEIR PROBABILITY

To begin with, let us assume that we wish to reliably communicate a single binary symbol $x$ through a binary symmetric channel of error probability $p$, as depicted in Fig. 6.1. We may repeat this symbol, say, 3 times. Then, we may consider that we send a 'word' of the form $xxx$, where $x$ belongs to the binary alphabet $\{0, 1\}$. At the receiving end, 8 possible words may correspond to the transmitted word $xxx$, depending on the error pattern $\underline{e}$ which occurred due to the channel noise. The components of $\underline{e}$ are independent binary variables due to the assumption that the channel is memoryless. The possible error patterns are listed together with their probability in the following table, where $\bar{x}$ denotes the binary complement of $x$, $\bar{x} = x \oplus 1$.

---

[3]In earlier works, and especially in [7], we named this quantity 'algebraic value' for translating the French 'valeur relative' used in [6], just meaning that it is endowed with a sign; it has however no relationship with algebraic codes and we hope that the word 'real' helps avoiding misunderstanding.

| received word | error pattern | probability |
|:---:|:---:|:---:|
| $xxx$ | 000 | $(1-p)^3$ |
| $\bar{x}xx$ | 100 | $p(1-p)^2$ |
| $x\bar{x}x$ | 010 | $p(1-p)^2$ |
| $xx\bar{x}$ | 001 | $p(1-p)^2$ |
| $\bar{x}\bar{x}x$ | 110 | $p^2(1-p)$ |
| $\bar{x}x\bar{x}$ | 101 | $p^2(1-p)$ |
| $x\bar{x}\bar{x}$ | 011 | $p^2(1-p)$ |
| $\bar{x}\bar{x}\bar{x}$ | 111 | $p^3$ |

One immediately notices that the probabilities do not depend on the transmitted symbol $x$, but only on the *weight* of the error pattern $\underline{e}$, defined as the number of symbols '1' it contains. An obvious generalization to $n$-fold repetition is that the probability of *a given* error pattern $\underline{e}$ of weight $w(\underline{e})$ equals

$$P_{\underline{e}} = p^{w(\underline{e})}(1-p)^{n-w(\underline{e})} \ . \tag{6.4}$$

Assuming the channel error probability $p$ to be less than $1/2$, the probability $P_{\underline{e}}$ of an error pattern of weight $w(\underline{e})$ is a decreasing function of $w(\underline{e})$ for any codeword length $n$. This conclusion holds, of course, independently of the transmitted symbol.

We also notice that the probability of occurrence of *any* error pattern of weight $w$ equals

$$P_w = \binom{n}{w} p^w (1-p)^{n-w} \tag{6.5}$$

since there are $\binom{n}{w}$ distinct possible error patterns of weight $w$, each of which of probability given by Eq. (6.4). This is a Bernoulli distribution with expected value (or mean) $np$ and variance (i.e., the expected value of the square of the difference between its actual value and its mean) $np(1-p)$.

## 6.3.2 DECISION ON A REPEATED SYMBOL BY MAJORITY VOTING

Let us go back to the case of a triply repeated bit $x$, and assume that the real values of the received bits are not available. Their common probability of error is denoted by $p$. If the received word consists of nonidentical bits, for instance 001, it is *sure* that a nonzero error pattern occurred since the transmitted word consisted of a same repeated symbol. On the other hand, if we receive a word made of a repeated bit, say 000, we *do not know* whether 000 was actually transmitted without error or if the actually transmitted word was 111 but the error pattern 111 occurred, transforming 111 into 000. We can say only that 000 is more likely to have actually been transmitted, since the probability of the error pattern 000, $(1-p)^3$, is larger than the probability $p^3$ of the error pattern 111, because $p < 1/2$. Therefore, the best assumption we can make is that 000 was transmitted. Similarly, the probability of the error pattern 001, $p(1-p)^2$, is larger than that of 110 which is $p^2(1-p)$, so the best assumption when 001 is received is that the error pattern 001 actually occurred so that the transmitted word was 000. In both cases, the best decision $\hat{x}$ as regards the transmitted bit is $\hat{x} = 0$.

We may thus formulate the optimum decision rule: the best assumption $\hat{x}$ as regards the transmitted bit is obtained by *majority voting*. For a symbol repeated $n$ times (instead of 3 times in the example), majority voting clearly defines a decision procedure if $n$ is odd. For $n$ even, a tie occurs if $n/2$ bits equal 1 (and the remaining $n/2$ equal 0). It is optimally solved by choosing the bit $\hat{x}$ at random (0 or 1 with probability 1/2). This optimum decision rule has as *average* probability of error the probability that an error pattern occurs with weight larger than $n/2$, namely:

$$P_{n\mathrm{rep}} = \sum_{w=(n+1)/2}^{n} \binom{n}{w} p^w (1-p)^{n-w}, \; n \text{ odd}, \tag{6.6}$$

$$P_{n\mathrm{rep}} = \sum_{w=n/2+1}^{n} \binom{n}{w} p^w (1-p)^{n-w} + \frac{1}{2}\binom{n}{n/2} p^{n/2}(1-p)^{n/2}, \; n \text{ even}$$

as a consequence of Eq. (6.5).

We notice that no improvement of the error probability results from repetition if $n = 2$ since then $P_{2\mathrm{rep}} = p(1-p) + p^2 = p$. If $n > 2$, however, repetition at the transmitting end and majority voting at the receiving one always improve the error probability since $P_{n\mathrm{rep}} < p$, and moreover $P_{n\mathrm{rep}}$ is a decreasing function of $n$ which approaches 0 when $n$ approaches infinity. The lack of improvement of the decision error probability in the case of $n = 2$ does not mean that twofold repetition is always useless: when the received word is 01 or 10, we know that no reliable decision can be taken on the transmitted bit so we may think of this bit as being erased. This is better than taking a wrong decision and can be exploited, e.g., when twofold repetition is combined with some other coding process.

Using a decision rule which results in a symbol belonging to the alphabet of the transmitting end, like majority voting, is referred to as a *hard decision*. We know that it is the best possible decision, but no assessment of the probability that it is correct or wrong is available. For instance, in the case of $n$-fold repetition with $n$ even, a binary decision is taken with a probability of 1/2 to be wrong if a tie occurs, but the bit which results from this decision cannot be distinguished from those which result from decisions taken with a large margin, hence having a small probability of error.

### 6.3.3   SOFT DECISION ON A REPEATED SYMBOL

A different situation arises with the use of *soft decisions*. A soft decision consists (ideally) of assessing the probability of a symbol by taking into account measurements on the corresponding received physical quantity as well as the constraints this symbol possibly obeys. At the single-bit level, measuring the channel output voltage implies a soft decision since we know that it is related to the probability of an error according to Eqs. (6.2) and (6.3). When we consider $n$-fold repetition of a bit, the fact that the repeated bits are identical is a constraint which can be taken into account. One easily shows that the optimum decision rule in this case is:

$$\hat{a} = \sum_{i=1}^{n} a_i, \tag{6.7}$$

where $a_i$ is the real value (referred to as *a priori*) of the $i$-th individual received symbol and where $\hat{a}$ is the *a posteriori* real value of the decision on the repeated symbol, optimally taking account of the fact that $n$ identical symbols were transmitted. The extreme simplicity of this decision rule is the main motivation for using the real value formalism. Interestingly, replacing the *a priori* real value $a$ by the *a posteriori* one $\hat{a}$ in Eqs. (6.2) and (6.3) still expresses the relationship of the *a posteriori* real value $\hat{a}$ of the decision on the repeated symbol with the error probability of this decision. Therefore, the output of a decoder has the same form as its input and the same relation to the error probability, which enables cascading several decoders, or iterated decoding. By comparison, taking a hard decision implies the loss of the reliability information borne by the magnitude of the real value. Multiple successive hard decisions thus incur considerable information loss, which practically hinders the use of schemes combining several codes. It turns out, however, that these schemes are very efficient provided successive soft decisions are implied in their decoding.

## 6.4 ENCODING A FULL MESSAGE

### 6.4.1 INTRODUCTION

The trouble with $n$-fold repetition is that reliably communicating a *single* symbol needs that *many* symbols are sent. The *rate R*, defined as the ratio of the number of symbols actually communicated to the number of sent symbols is here $R = 1/n$. This scheme provides a vanishingly small probability of error when $n$ approaches infinity, but at the expense of a vanishingly small rate. In any realistic situation, the rate should be as large as possible. Is it possible to encode a message of length $k$ into a codeword of length $n$ so as to reduce the error probability of the message? Moreover, is it possible to let $k$ and $n$ together approach infinity so as to keep a nonzero value of the rate $R = k/n$, while at the same time the probability of a decoding error vanishes?

The researches on error-correcting codes were intended to answer both questions and they succeeded in finding adequate solutions. We shall first introduce a simple example where the mathematical formalism already discussed can be employed, and we shall then generalize.

### 6.4.2 A SIMPLE EXAMPLE

Given a codeword length $n$ and a message length $k$, with $n > k$, we assume that we associate a codeword of length $n$ to any message of length $k$, resulting in a set of words named $(n, k)$ *block code*. We first consider small values of $k$ and $n$. Let us assume for instance that we wish to communicate a message of $k = 3$ bits (referred to as *information bits*) using a word of length $n = 6$ bits. Let $u_1u_2u_3$ denote this message. Let us assume that we append to it $n - k = 3$ redundancy bits (also referred to as 'check bits') $r_4$, $r_5$ and $r_6$, so we obtain the codeword $u_1u_2u_3r_4r_5r_6$. The redundancy bits are computed in terms of the information bits according to the following rule:

$$
\begin{aligned}
r_4 &= u_1 \oplus u_2 , \\
r_5 &= u_2 \oplus u_3 , \\
r_6 &= u_3 \oplus u_1 .
\end{aligned}
$$

The addition modulo 2 denoted by $\oplus$ has already been defined in Sec. 6.2.2. Its properties enable writing these equalities in the form of the three *parity checks*:

$$
\begin{array}{rcll}
u_1 \oplus u_2 \oplus r_4 &=& 0, & (c_{12}) \\
u_2 \oplus u_3 \oplus r_5 &=& 0, & (c_{23}) \\
u_3 \oplus u_1 \oplus r_6 &=& 0, & (c_{31})
\end{array}
\tag{6.8}
$$

which express that the number of bits '1' in each of the left-hand sides is zero or even. The check on the information bits $u_i$ and $u_j$ is denoted by $c_{ij}$, $i, j$ = 1, 2 or 3. With respect to the case of mere repetition dealt with in Sec. 6.3, we can say that each redundancy bit repeats an information bit in combination with another one. This enables a higher rate, and is almost equivalent to the mere repetition if the probability that two bits are simultaneously in error is low enough, i.e., if the error probability $p$ is small enough.

Notice that the above encoding process immediately generalizes to an arbitrary number $k$ of information bits by associating a redundancy bit to each of the $\binom{k}{2} = k(k-1)/2$ pairs of information bits. A code of length $n = k(k+1)/2$ and minimum distance $k$ results, to be referred to as a 'pair-checking code'. Such a code is very redundant for large $k$. We'll denote it by $\mathcal{PC}(k)$, so the code taken as example is $\mathcal{PC}(3)$.

A matrix formalism is convenient for describing this code. Let us represent the message by the row matrix (vector) $\underline{u} = [u_1 \, u_2 \, u_3]$ and the encoded word by the row matrix $\underline{c} = [u_1 \, u_2 \, u_3 \, r_4 \, r_5 \, r_6]$. Then the encoding rules can be expressed by the matrix relation (we assume that the entries of all matrices are dealt with as elements of the binary field, i.e., their addition is effected modulo 2):

$$
\underline{c} = \underline{u}G ,
\tag{6.9}
$$

where G is the 'generator matrix'

$$
G = \begin{bmatrix} 1 & 0 & 0 & 1 & 0 & 1 \\ 0 & 1 & 0 & 1 & 1 & 0 \\ 0 & 0 & 1 & 0 & 1 & 1 \end{bmatrix} .
\tag{6.10}
$$

Similarly, the set of parity checks can be summarized by the matrix relation:

$$
\underline{c}H^t = \underline{0} ,
\tag{6.11}
$$

where the superscript t denotes transposition, $\underline{c}$ denotes a codeword and $\underline{0}$ denotes the all-0 $(n - k)$-vector (row matrix), and where H is the 'parity-check matrix'

$$
H = \begin{bmatrix} 1 & 1 & 0 & 1 & 0 & 0 \\ 0 & 1 & 1 & 0 & 1 & 0 \\ 1 & 0 & 1 & 0 & 0 & 1 \end{bmatrix} .
\tag{6.12}
$$

Then, the matrix equality (6.9) tells how to generate a codeword in terms of the message to be transmitted, while the matrix equality (6.11) describes the constraints that any codeword $\underline{c}$

specifically obeys. The set $\mathcal{PC}(3)$ of words which correspond to all possible messages according to Eq. (6.9), which is also the set of words which obey constraints Eq. (6.11), constitutes what is referred to as a *linear code*. The term 'linear code' means that all codewords are obtained by linear combinations of the rows of the generator matrix, i.e., if $\underline{g}_1, \underline{g}_2, \ldots, \underline{g}_k$ are rows of G, then $\lambda_1 \underline{g}_1 \oplus \lambda_2 \underline{g}_2 \oplus \cdots \oplus \lambda_k \underline{g}_k$ is a codeword, where $\lambda_1, \lambda_2, \ldots, \lambda_k$ are elements of the binary field and the componentwise addition modulo 2 is denoted by $\oplus$. This implies that the all-0 word, which results from choosing $\lambda_1 = \lambda_2 = \cdots = \lambda_k = 0$, belongs to any linear code.

The table below lists all the codewords of the code $\mathcal{PC}(3)$ chosen as example. They are generated according to Eq. (6.9) with (6.10) as generator matrix and they obey the parity-check relation (6.11) with H given by (6.12).

| message | codeword |
|---------|----------|
| 000 | 000000 |
| 100 | 100101 |
| 010 | 010110 |
| 001 | 001011 |
| 110 | 110011 |
| 101 | 101110 |
| 011 | 011101 |
| 111 | 111000 |

The code has been designed so that the first three bits of any codeword are identical to the corresponding message. A code such that the encoded message appears in a given set of positions in its codewords is referred to as being in *systematic form*. Any linear code can be put in systematic form, and assuming it to be so does not incur any loss of generality as regards its performance.

Let us define the *Hamming distance* between two codewords as the number of bit positions where these words differ. The smallest Hamming distance between any two of the listed codewords, referred to as the *minimum distance* of this code, is 3. We do not need to check this property for any pair of codewords. It suffices to notice that the weight of the codewords (as defined in Sec. 6.2.2) is at least 3: if $\underline{c}_1$ and $\underline{c}_2$ belong to a linear code, then due to the linearity $\underline{c}_1 \oplus \underline{c}_2$ also belongs to this code, where the operation denoted by $\oplus$ has been defined above. The weight of $\underline{c}_1 \oplus \underline{c}_2$ equals the Hamming distance between $\underline{c}_1$ and $\underline{c}_2$ due to its very definition, but it is also the distance of the codeword $\underline{c}_1 \oplus \underline{c}_2$ to the codeword $\underline{0}$. Therefore, the set of Hamming distances between the words of a linear code reduces to the set of their weights.

Since the all-zero word $\underline{0}$ belongs to any linear code, no loss of generality results from assuming that the message to be transmitted is the sequence $\underline{0}$. The only 'ones' found in the output sequence then result from channel errors.

### 6.4.3 DECODING THE CODE TAKEN AS EXAMPLE USING THE SYNDROME

We first briefly discuss a decoding process of the code $\mathcal{PC}(3)$ taken as an example which is typical of the hard decoding of binary linear block codes. In such a hard-decision decoder the input as well as the output are symbols of the binary alphabet. We'll consider later 'soft-decision' decoders involving real values, which have significantly better performance at the expense of a more difficult implementation.

Let $\underline{c}'$ denote the word received when the word $\underline{c}$ is transmitted. We may write $\underline{c}' = \underline{c} \oplus \underline{e}$, where $\underline{e}$ is the error pattern. Let us compute the matrix product

$$\underline{s} = \underline{c}'H^t = \underline{e}H^t , \tag{6.13}$$

where the second equality is a consequence of Eq. (6.11). The $(n - k)$-vector $\underline{s}$ is referred to as the *syndrome*. If all its components are 0 ($\underline{s} = \underline{0}$), then $\underline{c}'$ belongs to $\mathcal{PC}(3)$ so the most likely assumption is that no errors occurred. Now, if an error pattern of weight 1 occurred, e.g., $\underline{e} = [1\,0\,0\,0\,0\,0]$, we have $\underline{s} = [1\,0\,1]$ i.e., the transpose of the first column of matrix (6.12). Similarly, had the error occurred on the $i$-th bit of $\underline{c}'$, the syndrome would be the transpose of the $i$-th column of matrix (6.12). Since all the columns of matrix H are distinct, the syndrome thus designates the column of matrix (6.12) at the position in $\underline{c}'$ where the error occurred, provided the error pattern is of weight 1. Since the alphabet is binary, a located error can immediately be corrected by replacing the corresponding bit by its complement. Of course, if an error pattern of weight larger than 1 occurs, the decoding rule above results in a wrong decision.

Computing the syndrome is the first step of the algebraic decoding of any linear code. In the above example, the relationship of the syndrome and the most likely error pattern is straightforward, because $\mathcal{PC}(3)$ was designed in order to obtain it. In more complicated codes, determining the most likely error pattern in terms of the syndrome is much more difficult. For short enough codes, a precomputed table can be used to associate the possible $2^{n-k}$ syndromes with the corresponding minimum-weight error patterns.

### 6.4.4    REPLICATION DECODING OF THE CODE TAKEN AS EXAMPLE

Let us now look at another means for decoding $\mathcal{PC}(3)$. We notice that the situation is not very different from the three-fold repetition of a single bit, except that the code rate is now 1/2 instead of 1/3. Indeed, we can solve the parity-check relations (6.8) in terms of $u_1$ when they contain it. Then, ($c_{12}$) results in $u_1 = u_2 \oplus r_4$ and ($c_{31}$) in $u_1 = u_3 \oplus r_6$. This means that both $u_2 \oplus r_4$ and $u_3 \oplus r_6$ repeat $u_1$ so it is possible to apply the above decision rules discussed in Sec. 6.3.3 to the 3 available received *replicas* of $u_1$: $u_1'$, $u_2' \oplus r_4'$ and $u_3' \oplus r_6'$, where the prime denotes the received symbols. We shall refer to the first replica as the 'trivial' one (it is the information symbol to be decoded itself) and the other two as 'compound replicas,' made of a combination of other symbols. The hard decision rule consists of majority voting among the above three replicas, and its result is the same as that of the decoding rule described in Sec. 6.4.3. We now consider soft decision decoding. The 'soft decision' rule consists of applying Eq. (6.7) to the corresponding *a priori* real values, namely $a_1$ which is associated with $u_1$, and the *a priori* real values associated with $u_2' \oplus r_4'$ and $u_3' \oplus r_6'$. One easily shows that the real value $a$ associated with the sum modulo 2 of two received binary symbols of real values $a_i$ and $a_j$ is such that

$$t(a) = t(a_i)t(a_j) \tag{6.14}$$

where $t(a)$ is defined as

$$t(a) = \tanh(a/2) = \frac{\exp(a) - 1}{\exp(a) + 1} .$$

Then this definition and Eq. (6.14) result in

$$a = \ln \frac{1 + t(a_i)t(a_j)}{1 - t(a_i)t(a_j)} = \ln \frac{\exp(a_i + a_j) + 1}{\exp(a_i) + \exp(a_j)} . \tag{6.15}$$

The real value associated with the sum modulo 2 of $\ell > 2$ binary variables is such that

$$t(a) = \prod_{i=1}^{\ell} [t(a_i)] , \tag{6.16}$$

hence its explicit expression is

$$a = \ln \frac{1 + \prod_{i=1}^{\ell} [t(a_i)]}{1 - \prod_{i=1}^{\ell} [t(a_i)]} . \tag{6.17}$$

Let $a_1, a_2, \ldots$ and $a_6$ denote the *a priori* real values of $u'_1, u'_2, u'_3, r'_4, r'_5$ and $r'_6$, respectively. Using Eq. (6.15), we can write the soft decision rule for optimally decoding $u_1$ as

$$\hat{a}_1 = a_1 + \ln \frac{1 + t(a_2)t(a_4)}{1 - t(a_2)t(a_4)} + \ln \frac{1 + t(a_3)t(a_6)}{1 - t(a_3)t(a_6)} , \tag{6.18}$$

where $\hat{a}_1$ is the *a posteriori* real value associated with $u_1$. Similarly, the decoding rules of $u_2$ and $u_3$ are

$$\hat{a}_2 = a_2 + \ln \frac{1 + t(a_3)t(a_5)}{1 - t(a_3)t(a_5)} + \ln \frac{1 + t(a_1)t(a_4)}{1 - t(a_1)t(a_4)} , \tag{6.19}$$

and

$$\hat{a}_3 = a_3 + \ln \frac{1 + t(a_2)t(a_5)}{1 - t(a_2)t(a_5)} + \ln \frac{1 + t(a_1)t(a_6)}{1 - t(a_1)t(a_6)} . \tag{6.20}$$

Figure 6.2 represents the encoder which generates the code taken as example in Sec. 6.4.2 if we interpret the square boxes as memories containing a single bit (1-bit memories) and the circles as modulo 2 adders. The 3 leftmost boxes are initially loaded with the 3 information bits $u_1, u_2,$ and $u_3$, while the 3 boxes labeled $r_4, r_5,$ and $r_6$ are initially left empty. The sums modulo 2 of the contents of the 1-bit memories which are connected to each modulo 2 adder are computed and load the 3 boxes labeled $r_4, r_5,$ and $r_6$. Then the six 1-bit memories contain a codeword.

Interestingly, Fig. 6.2 also represents a bit-by-bit decoder for the same code $\mathcal{PC}(3)$. We consider for instance decoding the information bit $u_1$. Let the 1-bit memories of the upper row be initially loaded with the received bits $u'_1, \ldots, r'_6$. We now assume that $c_{12}$ and $c_{31}$ compute

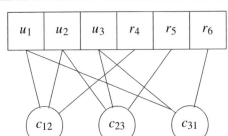

**Figure 6.2:** Implementation of the code defined by the parity-check equations [(6.8)].

the sum modulo 2 of the contents of the 1-bit memories connected to them, $u_1$ excepted. These sums modulo 2 are the two compound received replicas of $u_1$, so majority voting on $u'_1$, $u'_2 \oplus r'_4$ and $u'_3 \oplus r'_6$, where $u'_1$ is present in the first 1-bit memory while the other two are computed in the checks $c_{12}$ and $c_{31}$ and are applied to this 1-bit memory through the indicated connections. The decoded bit results from majority voting among $u'_1$, $u'_2 \oplus r'_4$ and $u'_3 \oplus r'_6$, and is denoted by $\hat{u}_1 = \text{maj}[u'_1, (u'_2 \oplus r'_4), (u'_3 \oplus r'_6)]$.

We have just described hard decision decoding of the code $\mathcal{PC}(3)$. The same figure also describes its soft decision decoding, provided we interpret the square boxes as containing real numbers and the circles as processors which compute the function defined by Eq. (6.15) in terms of the real quantities present at their inputs. The square boxes are initially loaded with the *a priori* real values $a_1$, $a_2, \ldots, a_6$ associated with bits $u_1, u_2, u_3, r_4, r_5$, and $r_6$, respectively, and the circular boxes $c_{12}$ and $c_{31}$ compute the real values associated with the 2 compound replicas of $u_1$. The real quantities present in the first square box and those coming from the circular boxes $c_{12}$ and $c_{31}$ are added, resulting in the soft decision (6.18) about $u_1$. The soft decision on the other two information bits are obtained similarly. A decoder like the one just described which both accepts and delivers soft decisions, i.e., real numbers the sign of which bears a binary decision and the magnitude of which measures the reliability of this very decision is referred to as a *soft-input soft-output*, or SISO, decoder.

The decoding procedures just described for the code $\mathcal{PC}(3)$ taken as example directly generalize to any pair-checking code $\mathcal{PC}(k)$ as defined in Sec. 6.4.2.

### 6.4.5   DESIGNING EASILY DECODABLE CODES: LOW-DENSITY PARITY CHECK CODES

Of the two operations of encoding and decoding, the latter is by far the most difficult. It is why it is expedient to design a code so as to make its decoding as easy as possible. To this end, let us now consider the generalization to much longer messages and codewords of the scheme of Fig. 6.2, as depicted in Fig. 6.3. The length of the row of 1-bit memories, i.e., the codelength $n$, is now much larger, say, 1000 or more. The row of $(n - k)$ parity checks is of proportionate size, say, 500 or so. Now, connect *at random* the 1-bit memories and the parity checks, only obeying the following conditions:

— each parity check is connected to at least 2 of the $k$ 1-bit memories which contain the information bits and to 1 of the $(n-k)$ ones which contain the check bits;

— the 1-bit memories connected to each parity check are few: they are at least 3 but do not need to be much more numerous (their number needs not be the same for all parity checks);

— each 1-bit memory is connected to at least one parity check.

The codes thus defined are referred to as *low density parity check* (LDPC) codes. They have been introduced and studied by Gallager [39]. Their presentation here is inspired by Berrou [23].

Just like the scheme of Fig. 6.2, that of Fig. 6.3 represents a decoder as well as an encoder. The hard decision decoder using majority voting has only a limited interest. Now replace the 1-bit memories with 1-real-number memories containing the real values of the corresponding received bits, and interpret the parity checks as processors acting on real numbers to compute real values of sums modulo 2 according to Eqs. (6.15) or (6.17). The real value thus computed in a particular parity-check processor is added to that already contained in the 1-real-number memory to which it is connected. We thus obtain a SISO decoder as defined in Sec. 6.4.4. Then a major fact is that decoding can be *iterated*. If the code rate $k/n$ is low enough with respect to the channel capacity, the decoding process described above results in an average improvement, the magnitudes being changed and the sign of certain real values (hence the corresponding binary decisions) being possibly inverted. Once these improved results have been obtained, the decoding process may be repeated in terms of the new real values and result in a further average improvement. The process can be iterated many times, until a stable enough result has been obtained. No theorem ensures that this process converges towards a well defined limit, but it actually does in practice. The performance thus obtained is close to that of turbocodes (see Sec. 6.7), i.e., small decoding error probabilities are obtained at rates close to the channel capacity.

Strictly speaking, this decoder is not optimal since decoding equations like Eqs. (6.18) to (6.20) are valid only provided the replicas of each information bit are disjoint, but no provision was made in order to ensure they are so. However, the scarcity of the bits combined in each parity check entails that only a small fraction of the replicas of information bits are not disjoint, so dealing with them as disjoint is an acceptable approximation which entails little impairment with respect to the optimal decoding.

Interestingly, although the low-density parity check codes are intended to make bit-by-bit decoding easy, their design also results in a distance distribution close to that which is obtained by random coding, as shown in [14].

Connecting in the scheme of Fig. 6.3 some of the information bits to many checks and other information bits to few checks, or even to none, results in an unequal-protection error-correcting code as the nested system to be discussed in Ch. 9. Moreover, it can be extended to provide an increased protection to already present information bits by including more information and check bits, increasing the number of parity checks and connecting the already present information bits to

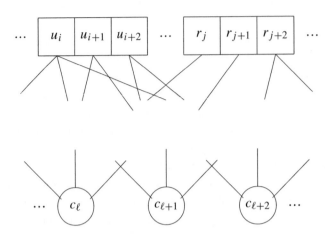

**Figure 6.3:** Scheme of a low-density parity check code.

the newly introduced parity checks, while keeping the already existing connections. This process is similar to that described in Sec. 9.1 to account for the setup of a nested system.

### 6.4.6    SOFT DECODING OF OTHER BLOCK CODES

The soft decoding rule based on replicas takes the simple form of Eqs. (6.18), (6.19), and (6.20) only if the replicas obtained by combining bits are disjoint in the sense that no bit belongs to more than one combination, or if they can be considered as approximately so. It turns out, unfortunately, that the codes such that disjoint replicas of the information symbols can be written are few, so this simple rule is far from being the general case. For nondisjoint replicas, a similar decoding rule can be written or implemented as an algorithm, but it is not as simple as in the above example. However, it can be always written as

$$\hat{a}_i = a_i + f_i(a_1, a_2, \ldots, a_{i-1}, a_{i+1}, \ldots, a_n) \tag{6.21}$$

where $\hat{a}_i$ denotes the *a posteriori* real value associated with the information bit $u_i$, $a_1, a_2, \ldots, a_n$ are the *a priori* real values associated with the $n$ received bits, and $f_i(\cdots)$ is a function of *a priori* real values other than $a_i$. We may refer to it as the *extrinsic* real value of $u_i$, i.e., contributed by all the other bits of the codeword. The explicit expression of this function in the general case is

$$f_i(a_1, a_2, \ldots, a_{i-1}, a_{i+1}, \ldots, a_n) = \ln \frac{\sum_{\underline{c} \in \mathcal{C}_{i0}} \exp(-\sum_j c_j a_j)}{\sum_{\underline{c}' \in \mathcal{C}_{i1}} \exp(-\sum_j c'_j a_j)} - a_i \,, \tag{6.22}$$

where $\mathcal{C}_{i0}$ (resp. $\mathcal{C}_{i1}$) denotes the set of words of the given code $\mathcal{C}$ having 0 (resp. 1) as $i$-th symbol, $\underline{c}$ denotes a codeword, $c_j$ its $j$-th bits, and $a_j$ denotes the corresponding *a priori* real value. This formula is rather complicated but fairly simple algorithms enable computing the function, at least for simple enough codes (we shall briefly describe some of these algorithms in Sec. 6.6.3). Moreover, approximations of $f_i(\cdots)$ can be computed with little performance degradation. The simplest one consists of keeping only the largest terms in the numerator and the denominator of the argument of the logarithm in Eq. (6.22), which results in:

$$f_i(a_1, a_2, \ldots, a_{i-1}, a_{i+1}, \ldots, a_n) \approx \min_{\underline{c} \in \mathcal{C}_{i1}} \sum_j c_j a_j - \min_{\underline{c}' \in \mathcal{C}_{i0}} \sum_j c'_j a_j - a_i . \qquad (6.23)$$

From the definition of $\mathcal{C}_{i0}$ and $\mathcal{C}_{i1}$, it is clear that $\hat{a}_i$ takes the form of the sum of $a_i$ and a term $f_i(a_1, a_2, \ldots, a_{i-1}, a_{i+1}, \ldots, a_n)$ which does not depend on $a_i$, since $\exp(-a_i)$ is a factor of the denominator of the fraction in the logarithm argument and its numerator does not depend on $a_i$. Therefore, Eqs. (6.22) and (6.23) may look trivial. Writing the decision rule (6.21) as the sum of $a_i$ and a term which does not depend on it is, however, meaningful due to the existence of algorithms which directly compute $f_i(a_1, a_2, \ldots, a_{i-1}, a_{i+1}, \ldots, a_n)$, exactly or approximately, in terms of real values other than $a_i$.

The search for block codes resulted in the invention of a large number of code families. The simplest one is that of the Hamming codes, which can correct a single bit in error. The code taken as example in Sec. 6.4.2 is actually a slightly modified Hamming code. Much more powerful and complex code families were designed using the algebra of finite fields, also referred to as Galois fields, collectively known as 'algebraic codes'. All of them were designed according to the *minimum distance criterion* for which a code is the better, the larger its minimum distance, and their decoding generally used the hard decision procedure of computing the syndrome and determining the most likely error pattern in terms of it. Although soft decoding using Eq. (6.21) is always possible for these codes (a means for doing so consists of applying the soft-output Viterbi algorithm or the BCJR algorithm to the trellis of a block code, see Sec. 6.6.3), computing the function $f_i(\cdots)$ given by Eq. (6.22) [even its approximate expression (6.23)] is quite complex for long codes and, moreover, soft decoding was seldom contemplated by the researchers on algebraic codes as involving probabilistic methods quite foreign to the algebra of finite fields, their discipline. We shall not discuss algebraic codes although they generated a plentiful literature, but consider instead the case of convolutional codes which has been fruitful in the design of practically efficient codes. Before doing so, we consider the more general framework provided by information theory.

# 6.5    ERROR-CORRECTING CODES IN THE FRAMEWORK OF INFORMATION THEORY

## 6.5.1    AN OUTLOOK ON THE FUNDAMENTAL THEOREM OF CHANNEL CODING

As yet, we have just introduced repetition as an error correction means and shown on a simple example that indirect repetition, i.e., involving combination of symbols by means of addition modulo 2, enables using less redundancy bits for a same number of replicas of each information bit. However, the sum modulo 2 of two or more unreliable bits is still less reliable than each of them, which means that the performance is degraded when the channel has a high probability of error (a low signal-to-noise ratio). We have seen also two very different decoding methods: hard decoding, where the received symbols are dealt with as symbols of the code alphabet, and soft decoding, where the *a priori* probabilities of the symbols are taken into account and where decoding moreover consists of reassessing the probability of the decoded symbols. More emphasis will be laid on the latter which is mandatory when several codes are combined (as in turbocodes to be considered later), in order to prevent a prohibitive performance degradation. The implementation of soft decoding is more complex than that of hard decoding, but the semi-conductor technology made such progresses since the invention of the transistor in 1948 that it is by now inexpensive and reliable.

Examples are necessarily simple, but efficient codes imply large codewords: a vanishingly small probability of decoding error can be obtained only asymptotically as the length of the codewords tends to infinity. Unlike the case of a symbol encoded by mere repetition, a good code should maintain a strictly positive rate $R = k/n$ when $n$ approaches infinity, where $k$ is the number of information bits, so $k$ also should approach infinity. That an errorless communication is asymptotically possible at any nonzero rate is actually a major result of information theory, referred to as the *fundamental theorem of channel coding* and already discussed in Sec. 5.3.1. It predated the attempts of building practical error-correcting codes (hence of finite length). Actually, this theorem acted as a strong incentive to the research of practical codes. The challenge was moreover that the proof of the theorem assumed *random coding*, the decoding of which is far too complex to be used.

Although the proof of the fundamental theorem of channel coding in its more general form involves rather difficult mathematics, we believe that it can be easily understood at the light of the *law of large numbers* in its weakest form, indeed a very fundamental concept. This law simply tells that the experimental outcomes have frequencies (defined as the ratio of the number of occurrences of given outcomes to the total number of trials) which approach with high probability the probabilities of the corresponding events when the number of trials approaches infinity. For instance, tossing a fair coin 10,000 times may result in the all-head (or all-tail) sequence, but with a probability[4] of only $2^{-10,000} \approx 10^{-3,010}$: a miraculous event! In strong contrast, the probability that the frequency of the heads or tails in the sequence thus obtained is close to 1/2 (say, in the range 0.47–0.53), is very large (see Eq. (6.25) in Sec. 6.5.3). Similarly, obtaining an error pattern of weight close to $np$ when $n$ symbols are transmitted over a binary symmetric channel of error probability $p$ has a large probability, provided the number of errors expected in the average, $np$, is large enough. So, although of course we do not know *where* the errors occur, we know that their *total number* is probably close to $np$ provided this number is large enough.

---

[4]This probability is actually that of *any given* sequence of that length. The difference is that the all-head sequence is unique, while there are $\binom{10,000}{5,000} \approx 7.98 \times 10^{3,007}$ sequences having exactly 5,000 heads and 5,000 tails.

## 6.5.2   A GEOMETRICAL INTERPRETATION

Let us consider the finite space of $n$-bit words, which has been referred to as the Hamming space of dimension $n$ and denoted by $S_n$ in Sec. 3.5.1. Its contains $2^n$ elements referred to as its points. In geometrical terms, knowing with high probability the total number of errors means that the received word is represented by a point which is probably close to the surface of an $n$-dimensional hypersphere of $S_n$ centered on the transmitted word, the radius of which is the expected number of errors $np$. If the minimum distance $d$ of the code is larger than twice this number, a point at the surface of this hypersphere is closer to the actually transmitted word than to any other codeword, hence unambiguously identifies it. Being with high probability close to this surface, the point which represents the received word is with high probability closer to the transmitted word than to any other codeword. The optimum decoding rule can thus be simply stated: *choose the codeword the closest to the received word*. The probability that this decoding rule results in an error is the smaller, the larger is $n$, and it vanishes as $n$ approaches infinity.

Again in geometrical terms, the design of the best code can be thought of as intended to *spread* $M < 2^n$ points belonging to $S_n$ so as to make them as far apart to each other as possible for the Hamming distance. For a given value of the error probability $p$ (assuming the channel to be binary symmetric), there is clearly a limit to the number $M$ of points which can be put in $S_n$ while keeping the distance between these points larger than $d = 2np$. Let $M_{\max}$ denote this number. The quantity

$$C = \lim_{n \to \infty} \frac{\log_2(M_{\max})}{n}$$

is the largest possible information rate which can be communicated through this channel. It is referred to as the *capacity* of the channel, expressed in shannons per bit, a quantity less than 1 since $M < 2^n$ is a necessary condition for the code to be redundant. It has been shown in Sec. 3.3.3 that the capacity of a binary symmetric channel of error probability $p$ is given by:

$$C_{\mathrm{bsc}} = 1 + p \log_2(p) + (1 - p) \log_2(1 - p) = 1 - \mathcal{H}_2(p) \text{ Sh/bit} .$$

The channel with binary input and real output, in the case of additive Gaussian noise, has a capacity significantly larger for the same value of the average error probability. It is expressed in terms of the signal-to-noise ratio $S/N$ by a more complicated formula which we do not give here.

## 6.5.3   DESIGNING GOOD ERROR-CORRECTING CODES
Random Coding

Random coding, i.e., the design of a code by choosing at random $M$ points into the $n$-dimensional Hamming space, is a means for regularly spreading points which is asymptotically optimum as $n$ approaches infinity. Random coding ensures that all the $n$ dimensions of the space $S_n$ are dealt with on an equal footing, in the average, and it results in achieving an average rate equal to the channel capacity. In the case of a code made of $M$ $n$-bit words, it consists of drawing at random each bit of a word independently of the other ones, with probability 1/2 of being 1 or 0, the $M$ words which make up the code being so drawn independently of each other. Then, the probability of a particular codeword $\underline{c}$ is $P_{\underline{c}} = 2^{-n}$, and the probability of obtaining any codeword of weight $w$ is:

$$P_w = \binom{n}{w} 2^{-n} . \tag{6.24}$$

The expected value, or mean, of this weight is $n/2$, and its variance, i.e., the expected value of the square of the difference between its actual value and its mean, is $n/4$. For very large $n$, a good approximation of this distribution is the Gaussian probability density: if we approximate $w/n$ by the continuous random variable $X$, the probability that $X$ belongs to the interval $(x, x + dx)$ is $p_X(x)dx$, where

$$p_X(x) = \sqrt{\frac{2n}{\pi}} \exp[-2n(x - 1/2)^2] . \tag{6.25}$$

This probability density function has a maximum at $x = 1/2$, hence for $w = n/2$, and assumes symmetrically decreasing values as $x$ differs from $1/2$. It is concentrated around $x = 1/2$, and the width of the region where it has nonneglible values decreases as $1/\sqrt{n}$, hence vanishes as $n$ approaches infinity.

Let us also mention that the fundamental theorem of channel coding has been proved by Gallager, assuming a randomly chosen error-correcting code [40].

Largest Minimum Distance of Error-Correcting Codes

No explicit means for designing a code with the largest possible minimum distance $d$ is known. However, asymptotically for $n$ approaching infinity, we know from random coding arguments that the largest possible minimum distance of a code is at least equal to the Gilbert-Varshamov bound $d_{GV}$, defined in the binary case by the implicit equation:

$$1 - k/n = \mathcal{H}_2(d_{GV}/n) , \tag{6.26}$$

where the binary entropy function $\mathcal{H}_2(p)$ has been defined in Eq. (3.6). For instance, if we assume as above that $k = n/2$, $d_{GV}$ is very close to $0.11 \times n$. Notice that the left-hand side of Eq. (6.26) measures the redundancy of the code and that $\mathcal{H}_2(\cdot)$ is an increasing function when its argument is smaller than $1/2$, so the larger the redundancy, the larger $d_{GV}$. For $k/n$ approaching 0, $d_{GV}$ tends to $n/2$.

For a nonbinary code, i.e., with $q > 2$, the Gilbert-Varshamov bound becomes:

$$1 - k/n = \mathcal{H}_q(d_{GV}/n) + (d_{GV}/n) \log_q(q - 1) , \tag{6.27}$$

with

$$\mathcal{H}_q(x) = -x \log_q(x) - (1 - x) \log_q(1 - x) = \frac{\mathcal{H}_2(x)}{\log_2(q)} .$$

Notice the presence in Eq. (6.27) of the additive term $(d_{GV}/n) \log_q(q - 1)$ which vanishes for $q = 2$. For very redundant codes $k/n$ vanishes. Then $d_{GV}/n$ approaches $(q - 1)/q$, a result easily deduced from Eq. (6.27).

Designing Effective Codes

Decoding a code obtained by random coding, however, is practically impossible, since decoding a single received word would imply to compare it with *all* possible codewords. Moreover, long codewords are necessary for good performance, which also means that the number of possible codewords, $2^{Rn}$, hence the number of needed comparisons, is extremely large for large $n$ (remember that $0 < R < 1$). It is why researches on error-correcting codes had to look for deterministic design rules, which enable their decoding with reasonably low complexity.

No general method is known for designing a code having $M_{\max}$ codewords, for an arbitrary value of $n$ and a given error probability $p$. Only a few coding schemes are known, or more often conjectured, to be optimal for specific values of $M$ and $n$, for some simple channels. In the absence of a general design rule, codes have been sought which are good for the minimum distance criterion, and the relevance of this simple criterion has not been questioned until the late 1980s. However, this criterion does not take into account the number of codewords at the minimum distance (or 'multiplicity') although a large value of the multiplicity results in performance impairment. The best codes known as yet, the turbocodes to be considered in more detail in Sec. 6.7, were not designed to comply with the minimum distance criterion. Their minimum distance is indeed often fairly small (at least with respect to the known bounds on the largest minimum distance) but their multiplicity is very small. These properties result in an 'error floor,' i.e., a decrease of the decoding error rate in terms of the signal-to-noise ratio much less steep when the signal-to-noise ratio is large than when it is small. Since the aim of an error-correcting code is to improve communication when the channel is poor, we may say that turbocodes are good where they are useful but bad where they are not.

We have seen that long words are needed for obtaining a small probability of decoding error. The design of block codes, their encoding and, above all, their decoding become complicated when an $n$-symbol codeword represents a message of $k$ information symbols where both $k$ and $n$ are large. There exists fortunately a family of linear codes which enable low complexity encoding, where only a small number of information symbols, say $k_0$, generate $n_0$ output symbols at a time. Repeating this encoding process $K$ times results in an arbitrarily long sequence of $n = Kn_0$ symbols as a function of $k = Kk_0$ information symbols: the codes thus obtained are referred to as *convolutional codes*. Besides the ease of their encoding, certain of these codes have very good weight properties, and their decoding is comparatively easy. Moreover, combining such codes according to efficient concatenation schemes and using adequate means for their decoding result in the most powerful error-correcting devices known as yet, the 'turbocodes', dealt with below in Sec. 6.7. The practical performance of turbocodes being close to the theoretical limit stated by information theory, i.e., the channel capacity, no significant improvement can be expected from other code families.

# 6.6    CONVOLUTIONAL CODES

## 6.6.1    CONVOLUTIONAL ENCODING

The *convolutional codes* are made of arbitrarily long codewords which can be generated by a simple encoding device. The simplest way for defining these codes consists of describing the encoder which generates them. They have no predetermined length, and their encoder consists of a device which receives the sequence of information bits and outputs the corresponding encoded sequence. For simplicity's sake, we assume that $k_0 = 1$, i.e., that the information bits enter the encoder one by one. Then there is no need to segment the input data stream into blocks, and the encoder is directly fed by the information bits. The heart of a convolutional encoder is a *shift register*, i.e., a set of a (generally small) number $m$ of binary memories, or cells, which are arranged in a row and connected to each other so that when a new bit enters the leftmost cell, the content of the memory of any cell is transferred into the cell immediately at its right. The content of the rightmost cell is lost unless it is explicitly connected to some other device. We shall represent a shift register according to Fig. 6.4. We assume that the binary symbols are represented by some physical quantity (e.g., a voltage, the polarity of which represents the binary symbol) which is maintained constant during some time interval $T$ referred to as 'bit duration,' after which the same physical quantity takes on the value which represents the next bit. For this reason, we interpret a memory cell as a delay element, the output of which is delayed with respect to its input by the bit duration $T$, and a shift register as made of a number of such delay elements.

Input    Output

**Figure 6.4:** Binary shift register of size $m = 3$. It consists of 3 binary memories, or cells. Each cell consists of a delay element of one-bit duration, represented by a D-shaped box. When a new bit enters the leftmost cell, the content of the memory of any inner cell is transferred into the cell immediately at its right; that of the rightmost cell is output.

A convolutional encoder is made of such a shift register which directly receives the information sequence. The contents of some of its memory cells are added modulo 2 to compute output bits on the one hand, and possibly 'recursion bits' which are fedback to the register input via a modulo 2 adder on the other hand. In the two examples to be presented, each encoder delivers two output bits for each input information bit. These two bits are generated in parallel, but they are most often transmitted in alternance (serially), each during a time interval $T/2$. The first example of a convolutional encoder is nonsystematic (the input bit is not output) and nonrecursive (there is no feedback) and has a memory $m = 2$ (Fig. 6.5). Despite its simplicity, it generates sequences which are at a Hamming distance of 5 to each other, hence they constitute a code which enables correcting all double errors. This convolutional code is a standard example in textbooks.

The encoder depicted in Fig. 6.6 is systematic since the input bit is directly output, and it also generates and outputs another bit, referred to as 'check bit.' The shift register and its connections will be referred to as a 'check bit generator' or 'rate-1 encoder.' That of Fig. 6.6 has been represented

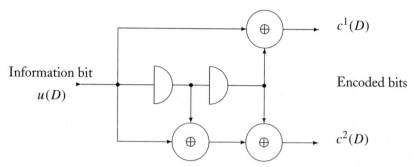

**Figure 6.5:** Example of a convolutional encoder of rate 1/2 with memory $m = 2$. For each entering information bit $u(D)$, two encoded bits $c^1(D)$ and $c^2(D)$ are output (the dependence on the dummy variable $D$ will be explicated later). Each point may assume one of two states, denoted by 0 and 1. The elements drawn as half-circles are one-bit delay operators (which together make a register), and $\oplus$ means addition modulo 2.

inside a dashed box. It uses a feedback and is thus referred to as 'recursive.' The feedback does not increase the minimum distance of a code, so it has no advantage in terms of the minimum distance criterion. However, it is very important in shaping the weight distribution of the code, since this feedback entails that the Hamming distance between most of the sequences it generates linearly increases with time. A code generated by such a recursive encoder is typical of a component code of a turbocode (see next section).

A very convenient formalism, referred to as the 'D-transform,' can be used to deal with shift registers and the sequences they generate instead of the vector notation used as yet for block codes. It uses polynomials and formal series of an indeterminate $D$ (which stands for 'delay,' meaning a delay of one bit duration with respect to some origin of time; $D$ is also referred to as 'delay operator'). We use a polynomial $N(D)$ in the indeterminate $D$ with coefficients belonging to the binary field to represent how the output is computed as the sum modulo 2 of bits contained in the register. The coefficients of $N(D)$ are 1 or 0, depending on the output of the memory cell corresponding to the degree of the indeterminate being connected or not to the modulo 2 adder which computes the encoder output. For instance, $N(D) = 1 + D^2 + D^3$ for the encoder depicted in Fig. 6.6. Similarly, a different polynomial $Q(D)$ specifies the recursion bit, where the coefficients are 1 or 0, depending on the output of the corresponding memory cell being used or not to compute the recursion bit fedback to the register input ($Q(D) = 1 + D + D^3$ for the encoder depicted in Fig. 6.6). We shall moreover assume that the degree of both polynomials $N(D)$ and $Q(D)$ equals the register memory $m$ and that both polynomials are irreducible, i.e., they cannot be expressed as a product of polynomials. Then it can be shown that the sequence generated by the encoder made of the shift register with its feedback determined by $Q(D)$ and its output computed according to $N(D)$, provided the initial state, i.e., the content of all the memory cells of the register, is 0 at the origin of time, is represented

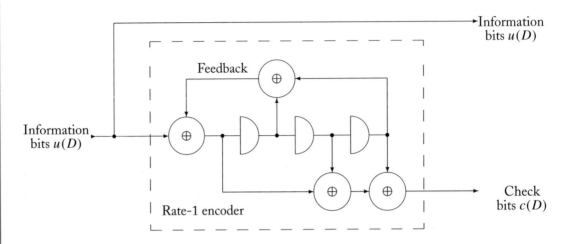

**Figure 6.6:** Example of a recursive systematic encoder of rate 1/2 with memory $m = 3$. Each entering information bit is output together with a check bit computed in the dashed box labeled 'rate-1 encoder.' This box contains the register and its connections. Each point may assume one of two states, denoted by 0 and 1. The elements drawn as half-circles are delay operators (of an information bit duration); they form together a register, and $\oplus$ denotes addition modulo 2. This encoder is referred to as 'recursive' because it involves a feedback.

as:

$$c(D) = u(D)\frac{N(D)}{Q(D)} , \qquad (6.28)$$

where $u(D)$ represents the input (information sequence) and $c(D)$ is the output sequence. For instance, if $u(D) = 1$, i.e., if the input sequence consists of a single 1 at the origin of time, the encoded sequence is represented as

$$c(D) = \frac{N(D)}{Q(D)} ,$$

which, since $Q(D)$ does not divide $N(D)$, can be expanded in a formal series with infinitely many terms, which thus represents an infinitely long sequence. For instance, we have in the above example:

$$c(D) = \frac{1 + D^2 + D^3}{1 + D + D^3} .$$

The reciprocal of the polynomial $Q(D)$ is

$$\frac{1}{1 + D + D^3} = 1 + D + D^2 + D^4 + D^7 + D^8 + D^9 + D^{11} + D^{14} + \cdots$$

where the right-hand side is periodic (the period here is the largest possible, $7 = 2^3 - 1$; the largest possible period for an $m$-cell shift register is $2^m - 1$), so

$$c(D) = \frac{1 + D^2 + D^3}{1 + D + D^3}$$
$$= 1 + D + D^4 + D^5 + D^6 + D^8 + D^{11} + D^{12} + D^{13} + D^{15} + D^{18} + D^{19} + D^{20} + \cdots$$

which is also periodic with period 7 except for the first two terms.

The requirement that the initial state of the shift register be zero implies that means for controlling the initial state are employed. For instance, a termination made of $m$ properly chosen bits may be appended to an $N$-bit information message so as to make the register content return to the zero state. Another solution to the register initialization problem consists of choosing a nonzero initial state such that the same state is reached after an $N$-bit message has been fed to the register, resulting in a 'tail-biting' convolutional code. The initial state needs then to be pre-computed in terms of the message. This is an elegant solution to the problem of initializing both component encoders of a turbocode (see next section).

Going back to the sequences generated by a convolutional encoder, we first notice a very important property: when an encoder uses feedback, i.e., is recursive, a finite-weight input sequence can generate an infinite-weight output sequence. This occurs for any input polynomial $u(D)$ which is not a multiple of the denominator $Q(D)$, i.e., for any input $u(D)$ such that $u(D)N(D)/Q(D)$ does not reduce to a polynomial, for instance if the input sequence is made of a single bit 1. On the other hand, in an encoder without feedback, any finite-weight input sequence would generate a finite-weight output sequence.

The multiples of the polynomial in denominator $Q(D)$, of degree $m$, constitute a fraction of $2^{-m}$ among all polynomials of any degree greater than or equal to $m$. This means that only a fraction $2^{-m}$ of all possible inputs to the encoder are encoded into sequences represented by polynomials and hence of finite weight, while all the others are represented by a formal series hence by an infinite-weight codeword. One could think of using a register of very long length $m$ so as to ensure that almost all generated encoded sequences have infinite weight, resulting in an almost errorless communication. Unfortunately, the complexity of decoding varies as $2^m$ (at least if optimum decoding is sought), which severely limits the possible values of $m$. Excellent weight properties can however be obtained if several short encoders are combined according to the *turbocode* scheme, which moreover can be almost optimally decoded with fairly low complexity. Performance close to the theoretical limit can indeed be obtained by this means with only two combined registers. Before we start describing the turbocodes, let us consider the crucial question of decoding convolutional codes.

## 6.6.2 SYSTEMATIC CONVOLUTIONAL CODES AND THEIR DECODING

Given the output sequence $c(D)$ given by Eq. (6.28), how can we decode it, i.e., recover the input sequence $u(D)$? Inverting this formula seems to be an answer. Doing so, we formally obtain:

$$u(D) = c(D)\frac{Q(D)}{N(D)} . \tag{6.29}$$

However, this relation fails to provide adequate means to decode $c(D)$. Remember that encoding is intended to protect the information message against errors. Applying Eq. (6.29) works only if no error occurred. What is actually received in the presence of errors is $c(D) \oplus e(D)$, where $e(D)$ is the $D$-transform representing the error pattern. Trying to use Eq. (6.29) with $c(D) \oplus e(D)$ instead of $c(D)$ results in $u(D) \oplus e(D)Q(D)/N(D)$. Except for the no-error case $e(D) = 0$ and a fraction $2^{-m}$ of all nonzero error patterns, $e(D)Q(D)/N(D)$ is a sequence of infinite weight. Instead of correcting an error, using Eq. (6.29) severely increases the number of errors, since a simple error pattern would generally result in the 'decoded' sequence having a larger weight error pattern than $e(D)$, often an infinite-weight one, for instance in the simplest case of a channel error pattern of weight 1. Most channel-error patterns would have such a catastrophic effect on the recovered sequence.

The set of check sequences $c(D)$ generated by a recursive encoder (i.e., involving feedback) does not actually constitute an error-correcting code. Remember that we have already stated that redundancy is a necessary condition of error correction. Since one bit of $c(D)$ is output every time an information bit (represented by a single term in $u(D)$) enters the register, there is no redundancy if we transmit only $c(D)$. We must actually transmit more mutually dependent bits, for instance both $u(D)$ and $c(D)$. Then, since two bits are transmitted every time an information bit enters the register, we have a redundant code of rate 1/2. Other rates can be obtained if information bits enter the register $k_0$ at a time and result in $n_0$ output bits at a time, with $n_0 > k_0$, so as to generate a redundant code of rate $R = k_0/n_0$. If the output contains the information bits at some specified places, the encoder is said to be 'systematic,' and it can be shown that no loss of generality results from considering only codes in systematic form. The output bits other than the information bits are referred to as 'check' or 'redundancy bits.' We may think of a systematic convolutional code, say of rate 1/2, as a kind of twofold repetition of the information bit where one of the repeated bits is transmitted while the other one is combined with others due to the encoder operation. The information bits and the check bits are most often transmitted according to a time-invariant pattern (e.g., in alternation for a 1/2-rate encoder), but what is important is that bits of both kinds are actually transmitted. Assuming from now on the code to be systematic, we shall denote any of its codewords by $[u(D), c(D)]$. At the decoding end, the systematic and check bits need to be distinguished, i.e., their location in the assumed time-invariant pattern should be known. This pattern is the exact equivalent of the 'reading frame' for properly identifying the three-nucleotide codons in the DNA molecule.

Then, the decoding devices or algorithms should not merely implement an inversion rule like (6.29), but must (ideally) determine the sequence which belongs to the code, (i.e., which can be generated by the encoder as described), say $[\hat{u}(D), \hat{c}(D)]$, the closest for the Hamming metric to the received (hence possibly erroneous) sequence, say $[u'(D), c'(D)]$ where $u'(D) = u(D) \oplus e_u(D)$ and $c'(D) = c(D) \oplus e_c(D)$ denote the sequence of received information and check bits, respectively, $e_u(D)$ and $e_c(D)$ denoting the error patterns which affect the information and check sequences, respectively. We may think of the process which leads to determine $[\hat{u}(D), \hat{c}(D)]$ as a means for *regenerating* the transmitted codeword $[u(D), c(D)]$. If it is successful, and it is almost always so if the code design is well fitted to the statistical characteristics of the channel errors, we may say that the effect of these errors is cancelled. In other words, thanks to its association with the check sequence, the information sequence has been made *resilient* to the channel errors.

We now discuss how convolutional codes are decoded. The main tool for describing decoding algorithms is the trellis diagram considered in the next section. A trellis diagram can also be drawn for block codes, so the same algorithms still work for block codes. We shall not describe the algorithms in detail, but only indicate their general principle.

### 6.6.3   THE TRELLIS DIAGRAM AND ITS USE FOR DECODING

Trellis of a Convolutional Code

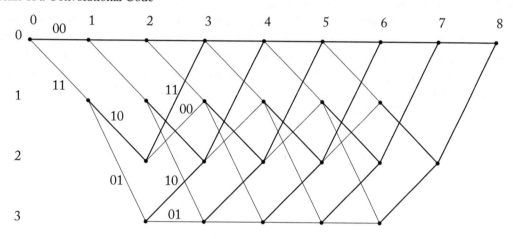

**Figure 6.7:** Trellis diagram corresponding to the encoder of Fig. 6.5. The states are the binary representations of the numbers from 0 to 3, plotted in terms of the step number. The branches drawn as thick lines correspond to information bits '0,' the thin ones to '1.' The labels of the branches are the bits output when the corresponding transition occurs. They have only been written for the first branches but periodically repeat themselves. Beyond the node level 6, we have drawn only the shortest paths going back to the zero state.

The trellis diagram graphically depicts the constraints that sequences generated by an encoder satisfy. Let us first consider the simple convolutional encoder represented in Fig. 6.5. How this encoder works can be graphically represented according to the trellis diagram of Fig. 6.7. The points in this figure represent the possible states of the encoder register at each step, i.e., every time a new information bit enters the encoder, and the lines between the points, referred to as 'branches,' represent the possible transitions from a state to another one. We assumed that both the initial and final states are 0. Apart for the first and last $m$ steps, the trellis is made of a repetitive pattern.

Trellis of a Block Code

A trellis can be similarly drawn for binary linear block codes. Let us consider, again, the binary linear block code $\mathcal{PC}(3)$ taken as an example in Sec. 6.4.2. Its parity-check matrix is given by (6.12) and is rewritten here for convenience:

$$H = \begin{bmatrix} 1 & 1 & 0 & 1 & 0 & 0 \\ 0 & 1 & 1 & 0 & 1 & 0 \\ 1 & 0 & 1 & 0 & 0 & 1 \end{bmatrix}.$$

One notices that all the columns of this matrix are binary representations of the numbers from 1 to 6. We define the state for a node of level $i$, which will play the same role as in the trellis of a convolutional code, as the number which has as binary representation the partial result of computing the product $\underline{c}H^t$ for the first $i$ symbols of the codeword $\underline{c}$. The state is plotted in terms of the node level in Fig. 6.8. If the first bit of $\underline{c}$ is '0,' the state at node level 1 is 0 (represented by vector [000]). If the first bit of $\underline{c}$ is '1,' one finds that the state at node level 1 is represented by the first row of $H^t$, namely [101], so the state is 5. If the second component is also '1,' the state obtained at node level 2 is represented by $[101] \oplus [110] = [011]$, hence is 3, etc. Only the paths which end at the state given by Eq. (6.11), i.e., 0, have been drawn as corresponding to possible codewords. The trellis thus obtained is represented in Fig. 6.8.

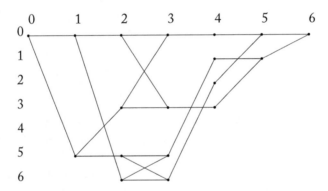

**Figure 6.8:** Trellis diagram corresponding to the block code of Sec. 6.4.2. The states are the numbers from 0 to 6, i.e., the binary representations of the partial result of computing the product $\underline{c}H^t$ for the first $i$ symbols of the codeword $\underline{c}$, plotted in terms of the step number. The horizontal lines correspond to an information bit 0, the oblique ones to an information bit 1.

Contrary to the trellis of a convolutional code, this diagram does not periodically repeat a same pattern. However, as expressing the coding constraints, it can be used as a support for decoding algorithms. These can be understood as means for computing the function $f_i(\cdots)$ in Eq. (6.21), using either the exact expression (6.22) or its approximation (6.23).

Decoding ultimately involves a decision which determines the most likely transmitted sequence among all the codewords. In the implementation of decoding or regeneration, however, it is more practical to proceed symbol by symbol, i.e., to decide what is the most likely symbol at each location in the sequence. Then the choice of an element among $q^k$ possible sequences is replaced by $n$ successive choices of a symbol among the $q$-ary alphabet (assuming an $(n, k)$ block code). Of course, each decision about a symbol is taken in terms of other symbols. An interesting possibility, which is especially exploited in turbocodes (see Sec. 6.7), consists of using previous provisional decisions to improve later ones, according to an iterated process (see Sec. 6.7.2).

A Decoding Algorithm Based on a Trellis: The Viterbi Algorithm

Let us first consider how the trellis diagram can be used for word-by-word hard decoding, i.e., determining the transmitted word the closest for the Hamming distance to the received one. Any transmitted word is represented as a path in the trellis, which depends on the input sequence of information bits. Examining a trellis shows that the encoder states being in finite number, there are as many paths which diverge from each node than paths which converge in it, namely two for a binary code (except for the beginning and end of the trellis). When two paths converge at a same node, their distance can be computed and one of them (ignoring a possible tie) is definitively better than the other. We may thus, at each node, keep only the best converging path, which is referred to as the *survivor*. The Viterbi algorithm consists of keeping only the survivor at each node of the trellis to which paths converge, an event which occurs for all nodes at each step after it has begun. After a sufficiently large number of steps, a single path survives and it corresponds to the best codeword which is thus decoded [74, 33].

Instead of using the Hamming metric, the distance can be computed between the paths so as to take account of the *a priori* probabilities of the received bits, resulting in 'soft-input decoding.' As regards the possibility of soft-output decoding, one can first notice that the distance between the paths which converge to a node measures the reliability of the decision on the corresponding information bit. Moreover, one can deal with the previous decisions as conditioned on the new ones, so the assessment of their reliability can be updated as the algorithms proceeds, resulting in symbol-by-symbol soft-input, soft-output decoding [9, 44].

Another Decoding Algorithm Based on a Trellis: The BCJR Algorithm

The Bahl, Cocke, Jelinek, and Raviv (BCJR) algorithm also uses the trellis diagram [3]. It consists of considering the reliability of a decision at each decoding step as conditioned on both the previous and future decisions (once a sequence is registered the future is available as well as the past). Then, the reliability of each symbol is evaluated starting from the knowledge of the initial state, based then on the previous states but still depending on the future ones. At the last step, the final state is known so the algorithm exactly determines the probability of the last symbol and can proceed in the reverse direction, providing the lacking information on the reliability of the future state at each decoding step. For this reason, this algorithm is also referred to as the 'forward-backward algorithm.'

## 6.7   TURBOCODES

### 6.7.1   DESCRIPTION AND PROPERTIES

The basic scheme for generating a turbocode involves the combination of two[5] convolutional systematic recursive encoders with an interleaver. Systematic recursive convolutional encoders have been defined in the previous section. An *interleaver* of length $N$ is a device having as input a sequence of length $N$, and an output which contains the same symbols as the input, hence also of length $N$, but in a different order. If its input sequence is $u_1, u_2, \ldots, u_N$, where $u_i$ is an element of some alphabet, for instance the binary one, its output is $u_{j_1}, u_{j_2}, \ldots, u_{j_N}$, meaning that the $j_i$-th element of the input sequence became the $i$-th in the output sequence. An interleaver $\Pi$ is completely defined by a set of $N$ different indices $\{j_i, 1 \le j_i \le N, i = 1, 2, \ldots, N\}$. For example, for $N = 7$, the interleaver defined by the set of indices $\{3, 6, 1, 2, 7, 5, 4\}$ transforms the word $0101100$ into $0001011$ and the word $1101011$ into $0111110$. The encoders currently used have a small memory (typically, $m = 3$) but, unlike the above example, the interleaver length $N$ is long (say, a few thousands of bits), so there is an extremely large number of possible interleavers, as many, i.e., $N!$, as the permutations of $N$ objects. When operating on some sequence $u(D)$ (represented by a polynomial of degree $N - 1$ in the indeterminate or delay operator $D$), the interleaver $\Pi$ transforms it into $\Pi[u(D)]$. The device which recovers $u(D)$ in terms of $\Pi[u(D)]$, often referred to as the de-interleaver, is another interleaver of length $N$, to be denoted by $\Pi^{-1}$, such that $\Pi^{-1}[\Pi[u(D)]] = u(D)$.

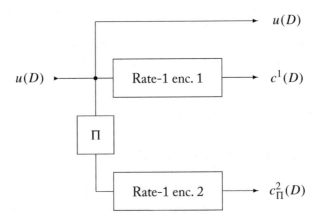

**Figure 6.9:**  Encoder of a turbocode of rate-1/3. The blocks labeled 'Rate-1 enc.' designate rate-1 encoders as represented by the dashed box in Fig. 6.6. The block labeled $\Pi$ represents the interleaver which implements permutation $\Pi$.

Then the encoder of a turbocode is organized according to Fig. 6.9: the input sequence of length $N$, represented by a polynomial in $D$ of degree $N - 1$, say $u(D)$, is output since the encoder

---

[5]Or more, but then some specific difficulties are met; two-component codes suffice for obtaining results close to the optimum.

is systematic. At the same time, each of two recursive rate-1 encoders like that of Fig. 6.6 computes check bits. These encoders may be identical to each other, or not. It will be convenient for us to distinguish them by the use of a superscript 1 or 2. One encoder is fed directly by the input sequence and the check sequence it generates is denoted by $c^1(D)$, while the other one is fed by the output of the interleaver, the input of which is the information sequence $u(D)$. The check sequence it generates is denoted by $c^2_\Pi(D)$. This basic scheme has inherently a rate of 1/3, but several tricks can be used so as to obtain higher rates (the most usual of them, referred to as 'puncturing,' consists of suppressing some of the generated bits according to a given periodic pattern). Considering the encoding scheme of a rate-1/3 turbocode will however suffice to understand the principle of turbocodes so we shall not consider other rates. We shall denote the output of this rate-1/3 encoder by $[u(D), c^1(D), c^2_\Pi(D)]$. It is a sequence of length $3N$, the weight of which is the sum of the weights of $u(D)$, $c^1(D)$ and $c^2_\Pi(D)$. This is a 'parallel' representation of the encoder output. In practice the encoder would generally involve a conversion into a 'serial' representation, consisting of successively transmitting the three binary outputs at three times the frequency of information bits.

An information sequence $u(D)$ made of a single bit '1' followed by '0's, i.e., of weight 1, results in both $c^1(D)$ and $c^2_\Pi(D)$ having a large weight (it would be infinite without the necessary truncation of the sequences to the interleaver length $N$, as we already noticed in Sec. 6.6.1). Then the overall output has large weight. If now we consider an information sequence of weight 2, we obtain $c^1(D)$ having a finite weight only if $u(D)$ is a multiple of the denominator $Q^1(D)$, which occurs if the two bits of the information sequence are located an integer number of periods apart in the sequence, i.e.,

$$u(D) = D^\alpha(1 + D^{\beta P}) \,, \qquad (6.30)$$

where $\alpha$ and $\beta$ are integers, and $P$ denotes the period of the formal series $1/Q^1(D)$. If $Q^1(D)$ is a 'primitive' polynomial[6], the content of the register successively represents in binary numeration all possible nonzero integers from 1 to $2^m - 1$ (in an order which depends on $Q^1(D)$) and we get a sequence of the longest possible period that can be generated by a register of length $m$, namely $P = 2^m - 1$. Then $Q^1(D)$ is a factor in the decomposition of $1 + D^{2^m-1}$ as a product of irreducible polynomials. It is so in the example of Fig. 6.6, where one easily checks that $1 + D^7 = (1 + D + D^3)(1 + D^2 + D^3)(1 + D)$ (do not forget that the coefficients are computed modulo 2), where both $1 + D + D^3$ and $1 + D^2 + D^3$ are primitive. Then the weight of the check sequence $c^1(D)$ corresponding to the information sequence (6.30) is $\beta$ times the weight of the periodic part of the generated sequence plus a small constant, hence it is the larger, the larger $\beta$. Similarly, $c^2_\Pi(D)$ of small weight is obtained if, and only if, $\Pi[u(D)]$ has the form (6.30) as above. It is possible to choose the permutation performed by the interleaver so that $u(D)$ and $\Pi[u(D)]$ do not simultaneously assume this form for small values of $\beta$. Then the weight of the generated sequence $[u(D), c^1(D), c^2_\Pi(D)]$ is large. For input sequences of larger weight, it may be very difficult to strictly avoid the simultaneous occurrence of $c^1(D)$ and $c^2_\Pi(D)$ having low weight, but the probability of

---

[6]A polynomial of degree $m$ is said to be primitive if taking the successive powers of one of its roots generates all the $2^m$ elements of a finite field.

such an event can be made very small. Although the design of a particular interleaver so as to obtain a specified weight performance is very hard, and although even the exact weight distribution which results from the use of a particular interleaver is difficult to compute, almost any interleaver will result in very few codewords of small weight and thus the overall performance will be good. Indeed, thanks to the interleaver operation and the fact that recursive convolutional encoders generate sequences of arbitrarily large weight, the set of sequences generated by a turbocode encoder resembles a set of sequences generated at random, so this encoding scheme appears as an approximate means for implementing random coding.

It turns out that a code generated at random is good with high probability. Indeed, random coding is the code construction that Shannon used in order to prove the fundamental theorem of channel coding (see Sec. 3.4.5) but, as we already mentioned, its decoding is practically impossible. The interest of the turbocode scheme is that its almost optimum decoding is comparatively easy, as we shall now show, while its weight distribution is close to that of random coding, hence intrinsically good.

### 6.7.2   SYMBOL-BY-SYMBOL SISO DECODING OF TURBOCODES

Let us first redefine the aim of decoding. Instead of trying to determine the codeword which has been transmitted with the highest probability, given the received word, let us extend the decoding role to determine the probability of each symbol of the code[7] (or at least of each of the bits in the information sequence) to be 0 or 1. Moreover, we may know the *a priori* probability of each of the symbols of the received word, as measured by the demodulator output (see Sec. 6.2.2). This kind of decoding is generally referred to as 'soft-input soft-output decoding' (SISO). We have already met SISO decoding in the very simple case of mere repetition, and in the simple example of Sec. 6.4.2. The general soft decision rule (6.21) was moreover of this kind. This redefinition of the decoding aim is an essential step for dealing with turbo decoding. It can be thought of as consisting of reassessing the probabilities of all the received bits, or at least of the information bits, given the *a priori* probabilities of all the received bits, so as to take account of the constraints of the code. The resulting probabilities will be referred to as *a posteriori* probabilities. We shall use the formalism of real values, as defined in Eq. (6.2), as we already did when dealing with mere repetition. Then, we may think of the decoding process as consisting of computing the *a posteriori* real value of each symbol (or of each information symbol) in terms of the *a priori* real values of all the received symbols. By *a priori* real value, we now mean the real value known at some present stage of the decoding process, prior to further processing, while *a posteriori* real value means that it resulted from taking into account some constraints yet ignored, or somehow improving the way constraints are accounted for.

Let us also notice that the *a posteriori* real value of any symbol, say $\hat{a}$, can be written as the sum of two terms, referred to as 'intrinsic' and 'extrinsic,' respectively. The first one, $a_{\text{in}}$, is the *a priori* real value of the symbol itself (which was referred to as the trivial replica in Sec. 6.4.4). The second

---

[7]The word made of the most probable bits is not necessarily identical to the codeword with the largest probability of having been transmitted. If the channel is not too bad, however, both almost always coincide.

one, $a_{ex}$, originates from the 'compound' replica obtained by expressing the same symbol in terms of others according to the constraints newly taken into account. The two replicas are independent as written in terms of disjoint sets of symbols. According to Eq. (6.21) but with different notation, the corresponding *a posteriori* real value is thus the sum of the intrinsic and extrinsic real values:

$$\hat{a} = a_{in} + a_{ex} . \tag{6.31}$$

We shall consider in the sequel elementary SISO decoders which implement the decoding rule (6.31) and are organized as in Fig. 6.10.

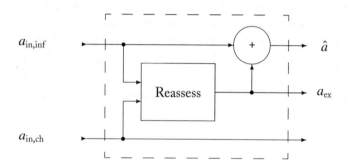

**Figure 6.10:** Elementary SISO decoder. Its two inputs are the *a priori* real values of the information bit and of the check bit, denoted by $a_{in,inf}$ and $a_{in,ch}$, respectively. Its three outputs are the *a posteriori* real value of the information bit $\hat{a}$, its extrinsic real value $a_{ex}$, and the *a priori* real value of the check bit which passes through the device without change. All these quantities are real and + denotes ordinary addition. The box labeled 'Reassess' implements the function $f_i(\cdots)$ in Eq. (6.21) so as to compute the extrinsic real value $a_{ex}$.

Let us now assume that we have an optimum (or nearly optimum) decoding rule for each of the two systematic codes (to be referred to as 'component codes') consisting of the two check sequences generated by the rate-1 encoders of the turbo encoder associated with the information sequence which, apart from the order of the symbols, is common to both. According to Eq. (6.31), we may write the *a posteriori* real value of any symbol which results from decoding the first code as:

$$\hat{a}^1 = a_{in} + a_{ex}^1 , \tag{6.32}$$

where $a_{ex}^1$ only takes account of the constraints of the first code. The same information symbol is present in the input sequence to the second encoder, at a location determined by the interleaver operation. We may thus consider the *a posteriori* real value from the first decoder (fed by the check sequence generated by the first rate-1 encoder) as the *a priori* real value of this symbol for the decoder which operates on the second component code, and write the decoding rule of this symbol with respect to the second code:

$$\hat{a}^2 = \hat{a}^1 + a_{ex}^2 = a_{in} + a_{ex}^1 + a_{ex}^2 , \tag{6.33}$$

where the extrinsic real value $a_{ex}^2$ is computed in terms of *a priori* real values consisting of the *a posteriori* real values delivered by the first decoder. The second equality results from Eq. (6.32).

A very important fact is now that the decoding process may be *iterated*. Indeed, once all the symbols have been decoded by the second decoder, we may think of the *a posteriori* real values from the second decoder as being improved estimates which can be used as *a priori* real values in the first one. However, the term $a_{ex}^1$ in the *a posteriori* real value as expressed by Eq. (6.33) has already been calculated when the first decoder was used, so it should be *subtracted* from the real values fed to it. Provided the extrinsic real value originating from a decoder is systematically subtracted[8] from the *a posteriori* real value from the other one used as input to this decoder, as shown in Fig. 6.10, the decoding process can in principle involve an arbitrary large number of iteration steps. It almost always converges, although the conditions for this to occur are very hard to express. In practice, the number of iteration steps has to be limited. Moreover, criteria for stopping the iteration as soon as a good enough result has been obtained can be used. The complex task of decoding the turbocode as a whole has been reduced to a succession of alternate decodings of the component codes, each of which is easily performed.

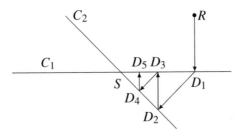

**Figure 6.11:** Geometric interpretation of the iterated decoding of two combined codes of a turbocode. Any point of the space represents the probability distribution of word components, initially conditioned on the received signals and then resulting from soft-output decodings. The point representing the optimally decoded codeword is denoted by $S$, the one which represents the received word by $R$. The loci of the points satisfying the constraints of each of the combined codes $C_1$ and $C_2$ are $C_1$ and $C_2$, respectively. By hypothesis, they both pass by $S$. The result of a first decoding which only takes account of $C_1$ is represented by $D_1$, that of the second decoding, which only takes account of $C_2$, by $D_2$, etc. The sequence of points $D_1$, $D_2$, $D_3$, ... tends to $S$ as the number of iteration steps increases.

The iteration process can be given a very simple picture in order to illustrate why it improves decoding (Fig. 6.11). We assume that we can define a space where a probability distribution is represented by a point and where the closeness of two distribution can be interpreted as the distance between the corresponding points (this can be done using the information-theoretic concept of 'cross-entropy,' also referred to as 'Kullback-Leibler divergence,' of two distributions). Then the

---

[8]Failing to do so would increase the magnitude of the *a posteriori* real value without improving its reliability; remember that the magnitude of a real value is intended to measure its reliability.

coding constraints can be represented by subspaces, say lines, and a probability distribution which satisfies the constraints of both codes is represented by the point where the corresponding lines intersect. The first decoding performed may be thought of as determining the point of the line which represents the constraint of the first code the closest to the one which represents the received point. To take account of the constraint due to the second encoding, we must start from this point and determine the point on the second line the closest to it. Since the second line only represents the constraint due to the second code, we must again determine the point on the first line the closest to the lastly obtained one, etc. After a sufficient number of iteration steps, a point close enough to the intersection of the two lines is reached and almost optimum decoding of the two combined codes has been performed. At variance with the figure, the two loci associated with the codes have more than a single intersecting point. There are indeed as many intersecting points as codewords, so the convergence indicated here is only a local property.

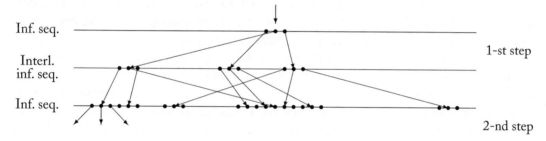

**Figure 6.12:** Diffusion of the dependence in the iterated decoding process. The lines labeled 'Inf. seq.' and 'Interl. inf. seq.' refer to the original information sequence and the one which results from its interleaving, respectively, at the 1-st iteration step and the 1-st half of the second one, from top to bottom. The points locate particular bits in the sequence. See the text for comments.

We can also illustrate how the iterated decoding proceeds in the information sequence with the help of Fig. 6.12. Remember that we interpreted the encoding as a kind of indirect repetition, i.e., where an information bit is repeated as combined modulo 2 with other information and check bits. Let us assume that this combination only concerns the two neighbors of each information bit (the combination also involves check bits which are not represented in the figure). The top horizontal line represents the information sequence at the beginning of the first iteration step. We consider the particular bit indicated by the vertical arrow at the top of the figure. As assumed, the encoding by the first component code made it dependent on its two neighbors. This means that, at the first half of the first iteration step (i.e., decoding in terms of the 1-st component code), the real value of the bit initially considered as well as those of its neighbors have been recomputed in terms of their own *a priori* real values and that of the corresponding checks bits. The *a posteriori* real values thus obtained are used as *a priori* real values for the second half of the first iteration step, which consists of decoding the interleaved information sequence in terms of the second component code. The arrows indicate the location of the interleaved bits. Due to the encoding by the second code, the neighbors of each of these bits in the interleaved information sequence have been made dependent and their *a posteriori* real values have been recomputed according to the second component code. At the first half of the second iteration step, the information bits are again ordered according to the original sequence, thanks to the de-interleaver operation. The bits initially considered have recovered their original place, but the neighbors of the corresponding bits in the interleaved sequence are located somewhere, generally far apart in the de-interleaved sequence. Iteration consists of repeating this process. Clearly, there is a diffusion of the dependence relationships through the entire information sequence, which eventually makes the *a posteriori* real value of an information bit depend on increasingly many other ones in an increasingly more complex fashion. We have a kind of 'diffusion of the dependence' as the decoding iteration proceeds. Remember that, in the average, the decoding process results in an increase of the magnitude of the *a posteriori* real values, hence of the reliability of the decoding decisions. Even if the improvement is small at the first iteration steps, a large number of steps eventually results in an almost sure decision if the code rate does not exceed the channel capacity. We may think of the decision involving each single bit as cumulating more and more information from the remainder of the sequence of received bits as the iteration of decoding proceeds.

As regards the implementation of the iterative decoding process, we may schematically think of cascading several decoding modules, each performing an iteration step. One of these modules is represented in Fig. 6.13. Its inputs can be connected to the outputs of the previous one, and its outputs to the inputs of the next one. Besides SISO decoders for each of the component codes, this elementary decoder involves an interleaver and a de-interleaver so as to ensure that both SISO decoders operate on the same information bit. The *a posteriori* real values of successive bits are correlated due to the dependency created by encoding and, in particular, the errors of each SISO decoder appear in bursts. The interleavers used in the decoding device spread out the *a posteriori* real values (and their possible errors occurring in a burst) before the sequence is fed to the other one, thus ensuring that the successive input *a priori* real values are uncorrelated, a condition for the validity of SISO decoding. This is a second important role of the interleaver, besides that of shaping the overall weight distribution of the turbocode which we already noticed.

### 6.7.3    VARIANTS AND COMMENTS

In the device represented in Fig. 6.13, we assumed that the *a priori* real values of the check bits pass through the elementary SISO decoders without change. As represented in Fig. 6.10, these decoders only update the *a posteriori* real values of the information bits. However, the formulas which express the *a posteriori* real value of any received bit in terms of the *a priori* real values of all the received bits, Eqs. (6.21) and (6.22), enable computing the *a posteriori* real values of the check bits, too. The choice of not doing so has the advantage of keeping samples of the channel output throughout the iterated decoding process, to the benefit of its stability, but schemes where both the information and check real values are updated can be contemplated.

The iterated decoding of turbocodes may actually be implemented quite differently from the scheme of Fig. 6.13 since the same device may be used several times at a speed greater than the input

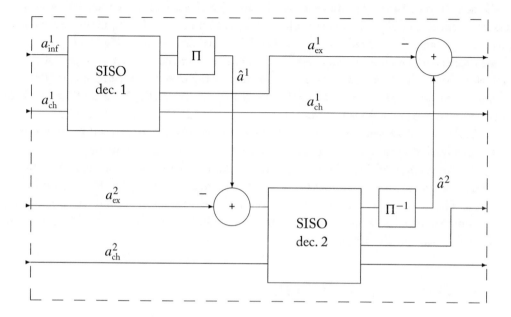

**Figure 6.13:** Decoding module (dashed box) for the turbocode of rate 1/3 generated by the encoder of Fig. 6.9. Iteration results from cascading a number of such modules. Contrary to Fig. 6.9, all the quantities considered are real: real values $a^i_{inf}$ and $a^i_{ch}$ of the information and check bits, respectively, and extrinsic real values as defined in the text, $a^i_{ex}$, with $i \in \{1, 2\}$. The boxes labeled 'SISO dec.' are identical to the elementary SISO decoder of Fig. 6.10. The extrinsic real value input is 0 in the first module, since then no extrinsic real value has yet been generated.

bit-rate. Then the iteration can be implemented using a single elementary decoder connected in a feedback loop.

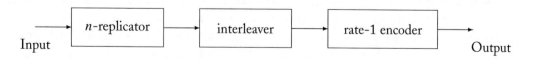

**Figure 6.14:** A fully serial schematic representation of a turbo encoder of rate $1/n$. The box named $n$-replicator designates a device which successively delivers $n$ times its input symbol. The interleaver scrambles the sequence it receives and the rate-1 encoder creates dependency between a number of successive symbols. Among the 3 devices, only the $n$-replicator generates redundancy.

We already stated that the information bits and the check bits, which appear as separate outputs in the encoder of Fig. 6.9, are generally transmitted 'serially,' i.e., in alternation according to a regular time pattern. We may redraw Fig. 6.9 in a fully serial fashion, as in Fig. 6.14, where three blocks are connected to each other: an *n*-replicator, an interleaver and a rate-1 encoder. Of course, each of these blocks results from a serial rearrangement of the elementary components (memory cells, connections, etc.) of the corresponding blocks of Fig. 6.9. Interestingly, each of the blocks of Fig. 6.14 performs one of the three functions which can be expected from a good encoder since they provide *redundancy, randomness,* and *mutual dependence,* respectively. We may thus think of Fig. 6.14 as describing a kind of paradigmatic encoder. Moreover, it turns out that some improvements can be obtained from variants of this scheme, especially if the number of copies generated by the *n*-replicator is made irregular, some symbols being repeated more than others while keeping the overall rate constant. The ease of decoding demands, however, that the decomposition of the overall code into short component codes be conserved, so as to enable splitting the whole task of decoding into alternate iterated decodings of simple codes.

## 6.8    HISTORICAL OUTLOOK

We now give a historical outlook on error-correcting codes, restricted to the descent of turbocodes. It is very sketchy and many other references could have been cited. Of course, the papers by Shannon are the starting point of the researches on error-correcting codes [72]. Pioneering works concerning block codes were due to Hamming and Golay in the late fourties [45, 42]. They originated in the algebra of finite fields and developed as 'algebraic coding', which is well accounted for in books like [65, 55], but has not been dealt with here. Convolutional codes were invented by Elias in 1955 [31]. The most important means for decoding convolutional codes has been the Viterbi algorithm [74, 33]. The trellis of a block code, enabling its SISO decoding, was introduced by Battail [8] and Wolf [78]. Battail [9] and Hagenauer and Hoeher [44] described a soft-output version of the Viterbi algorithm. Another SISO-type decoding algorithm was published earlier by Bahl et al. [3] and is now widely used. Replication decoding, by Battail et al. [6, 7], can be considered as a SISO-type version of Massey's threshold decoding [57]. Questioning the minimum distance criterion for the design of codes and suggesting that a good code should mimic random coding appeared in [10, 11], an idea which was later developed in [12]. The original paper on turbocodes by Berrou et al. dates back to 1993 [21]; turbocodes were given later a more comprehensive discussion in [22]. Also, see [43] for an excellent paper popularizing turbocodes. It was belatedly recognized that Gallager's 'low-density parity-check codes' [39] alluded to above in Sec. 6.4.5 are related to turbocodes as regards their decoding process and their performance. Since 1993, turbocodes were probably the topic which prompted the most important amount of work since the very beginning of researches on error-correcting codes: thousands of relevant bibliographic references about them are now available.

## 6.9    CONCLUSION

This chapter provided an introduction to error-correcting codes which hopefully enables under-standing how they work and why they are efficient. It has been restricted to a few important topics hence is very far from being exhaustive. It kept the mathematical apparatus to a minimum and relied on the reader's intuition to deal with the practically most successful codes. Also, the strong con-nection of error-correcting codes with basic concepts of information theory has been emphasized. Much of this text relies on rather unconventional concepts, vocabulary and notation elaborated by the author for many years. To some extent, they paved the way to the invention of turbocodes, which shows that their value is not merely subjective.

# Part III

# Necessity of Genomic Error Correcting Codes and its Consequences

CHAPTER 7

# DNA is an Ephemeral Memory

We now present computations of DNA capacity. Remember that information theory enables measuring the ability of any channel or memory element to convey information through space or time by its capacity, a quantity defined in Sec. 3.3.3 above. To compute it, we need to know what kind of events impair the basic information-bearing elements, namely, the nucleic-base pairs; and at which frequency these 'error events' occur. As regards the type of error to be considered, we already mentioned in Sec. 3.7 that it will be restricted to substitution and erasure, namely the cases where a wrong symbol replaces the correct one and where the actual symbol cannot be identified as belonging to the alphabet, respectively. In both cases, the length of the sequence remains unchanged, at variance with the cases of deletion or insertion where a symbol is removed from the sequence or inserted in it, resulting in its length decreasing or increasing by one. These latter events are not so uncommon in DNA, but it seems that their occurrence is less frequent than that of substitutions and erasures. Moreover, these cases are significantly more difficult to deal with so substitutions or erasures have been more thoroughly studied by information and channel coding theorists. To begin with, we compute the probability of symbol erasure or substitution as a function of time, assuming the errors are either substitutions or erasures, moreover occurring at a given constant frequency. We defer to Sec. 7.3 a closer examination of how the actual frequency of these error events can be estimated.

## 7.1 COMPUTING THE PROBABILITY OF SYMBOL ERASURE OR SUBSTITUTION AS A FUNCTION OF TIME

### 7.1.1 SYMBOL ERASURE PROBABILITY

We assume here that only errors of a single type occur, either substitution or erasure. We begin with erasure, the simplest case, assumed to occur with a constant frequency $\nu_{er}$, meaning that a symbol is erased during the infinitesimal time interval $dt$ with probability $\nu_{er}dt$. We first compute the probability $p_{er}(t)$ that a symbol has been erased at time $t \geq 0$. A symbol can be erased only if it has not been already erased, so we can express the probability $p_{er}(t + dt)$ as

$$p_{er}(t + dt) = p_{er}(t) + \nu_{er}dt[1 - p_{er}(t)]$$

which is equivalent to the differential equation

$$p'_{er}(t) = \nu_{er}[1 - p_{er}(t)]$$

where $p'_{er}(t)$ denotes the derivative of $p_{er}(t)$ with respect to time. Its solution satisfying the initial condition $p_{er}(0) = 0$ is

$$p_{er}(t) = 1 - \exp(-\nu_{er}t) . \tag{7.1}$$

The slope of the graph of $p_{er}(t)$ at the origin, $p'_{er}(0)$, equals $v_{er}$ and $p_{er}(t)$ tends to the horizontal asymptote $p_{er}(\infty) = 1$. Thus, after a long enough time the given symbol is eventually erased. Notice that this result does not depend on the alphabet size $q$.

## 7.1.2   SYMBOL SUBSTITUTION PROBABILITY

Consider a symbol from an alphabet of size $q$. In the case of DNA, $q = 4$ for the alphabet made of the nucleotides **A, T, G,** and **C**, or $q = 2$ if we only distinguish as symbols, for instance, their chemical structure purine, **R**, or pyrimidine, **Y**. We assume that this symbol incurs a substitution during the infinitesimal time interval $dt$ with probability $v_{su}dt$, where $v_{su}$ is constant. We moreover assume that the symbol which is substituted for the initial one is randomly chosen with probability $1/(q - 1)$ among the other $q - 1$ symbols of the alphabet (this is the worst probability distribution in case of error, according to information theory).

We now compute the probability of error of a given symbol as a function of time. Let $p_{su}(t)$ denote the probability that this symbol differs from the initial (correct) one at time $t \geq 0$. The given symbol is identical to the initial one with probability $1 - p_{su}(t)$, and in this case the probability of error increases during the interval $(t, t + dt)$ by an amount of $v_{su}dt$. But if the given symbol is already in error, an event of probability $p_{su}(t)$, the probability of error *decreases* by an amount of $v_{su}dt/(q - 1)$ since the newly occurring error can recover by chance the initial symbol. We can thus express the probability $p_{su}(t + dt)$ as

$$p_{su}(t + dt) = p_{su}(t) + v_{su}dt[1 - p_{su}(t)] - v_{su}dt\frac{p_{su}(t)}{q - 1} = p_{su}(t) + v_{su}dt\left[1 - \frac{q}{q - 1}p_{su}(t)\right],$$

which is equivalent to the differential equation

$$p'_{su}(t) = v_{su}\left[1 - \frac{q}{q - 1}p_{su}(t)\right], \tag{7.2}$$

where $p'_{su}(t)$ denotes the derivative of $p_{su}(t)$ with respect to time. Its solution satisfying the initial condition $p_{su}(0) = 0$ is

$$p_{su}(t) = \frac{q - 1}{q}\left[1 - \exp\left(-\frac{q}{q - 1}v_{su}t\right)\right]. \tag{7.3}$$

The slope of the graph of $p_{su}(t)$ at the origin, $p'_{su}(0)$, equals $v_{su}$ and $p_{su}(t)$ tends to the horizontal asymptote $p_{su}(\infty) = (q - 1)/q$. This asymptotic behavior for $t$ approaching infinity means that after a long enough time the given symbol no longer depends on the initial one and becomes random with uniform probability over the alphabet.

## 7.2   CAPACITY COMPUTATIONS

### 7.2.1   CAPACITY COMPUTATIONS, SINGLE-STRAND DNA

Assuming the errors to be all of the same type (either erasures or substitutions) it is a simple matter to compute the capacity of DNA in terms of the error probabilities (7.1) and (7.3) just obtained. The channels are symmetric in both cases, so their capacity equals the mutual information $I(X; Y) = H(Y) - H(Y|X)$ for equiprobable input symbols.

The capacity of the erasure channel has already been computed in Sec. 3.3.3, namely, $p_{er}$ denoting the erasure probability:

$$C_{q,er} = [1 - p_{er}] \log_2 q \tag{7.4}$$

shannons per symbol hence, letting $p_{er} = p_{er}(t)$ according to Eq. (7.1):

$$C_{q,er} = \exp(-\nu_{er} t) \log_2 q , \tag{7.5}$$

where the subscript indicates the alphabet size $q$ and the nature of the assumed error, 'er' standing for 'erasure.'

Similarly, if we assume in the case of $q$-ary symbols that all errors consist of substitutions of the type specified above, occurring with probability $p_{su}$, we obtain after straightforward computations the capacity:

$$C_{q,su} = \log_2 q - \mathcal{H}_2(p_{su}) - p_{su} \log_2(q - 1) \tag{7.6}$$

shannons per symbol, where the binary entropy function has been defined as $\mathcal{H}_2(p) = -p \log_2(p) - (1 - p) \log_2(1 - p)$. The subscript is similarly intended to indicate the size $q$ of the alphabet and 'su' stands for 'substitution.' Letting in the above expression $p_{su} = p_{su}(t)$ as given by Eq. (7.3) results in

$$
\begin{aligned}
C_{q,su}(t) = {} & \frac{q-1}{q}\left[1 - \exp\left(-\frac{q}{q-1}\nu_{su} t\right)\right] \log_2\left[1 - \exp\left(-\frac{q}{q-1}\nu_{su} t\right)\right] \\
& + \frac{1}{q}\left[1 + (q-1)\exp\left(-\frac{q}{q-1}\nu_{su} t\right)\right] \log_2\left[1 + (q-1)\exp\left(-\frac{q}{q-1}\nu_{su} t\right)\right] ,
\end{aligned}
\tag{7.7}
$$

which expresses the capacity of the genomic channel as a function of time in the presence of a constant substitution frequency $\nu_{su}$.

Notice that the error probabilities $p_{er}(t)$ and $p_{su}(t)$ in Eqs. (7.1) and (7.3), hence the capacities $C_{q,er}(t)$ and $C_{q,su}(t)$ in Eqs. (7.5) and (7.7), depend on time through the product $\tau = \nu t$ ($\nu$ stands here for either $\nu_{er}$ or $\nu_{su}$), a dimensionless quantity, which can be interpreted as a measure of time using $1/\nu$ as unit, to be referred to as *time constant*. The formulas (7.5) and (7.7) account for the degradation of the channel capacity due to the accumulated errors. The capacities decrease from $\log_2 q$ for $\tau = 0$, with a slope equal to $-\nu_{er}$ for erasures and $-\infty$ for substitutions, down to 0, *exponentially* for $\tau$ approaching infinity.

For the purpose of illustration, let us assume that the relevant alphabet is binary (say, {R,Y}, i.e., only distinguishes the purine or pyrimidine chemical structure of a nucleotide). The capacity given by Eq. (7.7) for $q = 2$ is:

$$
\begin{aligned}
C_{2,su}(t) = {} & \frac{1}{2}\{[1 - \exp(-2\nu_{su} t)] \log_2[1 - \exp(-2\nu_{su} t)] \\
& + [1 + \exp(-2\nu_{su} t)] \log_2[1 + \exp(-2\nu_{su} t)]\} .
\end{aligned}
\tag{7.8}
$$

Similar expressions would be obtained with other alphabet sizes.

### 7.2.2   CAPACITY COMPUTATIONS, DOUBLE-STRAND DNA

The capacities just computed are only relevant to single-strand DNA. However, the availability of pairs of nucleotides as elements of the alphabet enhances the capacity.

In the presence of erasures, a receiver dealing with pairs of nucleotides as elements of the alphabet will deal with a nucleic-base pair as erased only if its nucleic bases are both erased. Assuming the erasures of each base in the pair to be mutually independent events results in the pair erasure probability being the square of the erasure probability of a single base, so $p_{er}^2(t)$ should be substituted for $p_{er}(t)$ in Eq. (7.4). The capacity thus obtained is:

$$C_{q,er,ds} = \exp(-\nu_{er}t)[2 - \exp(-\nu_{er}t)]\log_2 q , \qquad (7.9)$$

where the appended subscript 'ds' stands for 'double strand.' It has been plotted in terms of $\tau = \nu_{su}t$ in Fig. 3.11.

In the case of substitution, an error can affect a single base or both. A substitution which affects a single base necessarily destroys the pair complementarity, so it must be interpreted as an erasure of the pair. Substitutions in both nucleic bases result in either a substitution or an erasure of the pair: a substitution when they result in a complementary pair, an erasure when they do not.

The case $q = 4$ is rather complicated and is not developed here. Assuming again for the purpose of illustration that the alphabet is binary, e.g., {**R,Y**}, a straightforward computation of the capacity in the presence of substitutions, taking account of the availability of pairs of nucleotides, results in:

$$C_{2,su,ds} = 1 + \mathcal{H}_2[2p_{su}(1 - p_{su})] - 2\mathcal{H}_2(p_{su}) .$$

Replacing $p_{su}$ with its expression in terms of $\nu_{su}t$ according to Eq. (7.3) results in:

$$
\begin{aligned}
C_{2,su,ds}(t) \;=\; & \frac{1}{2}\{[1 - \exp(-2\nu_{su}t)]^2 \log_2[1 - \exp(-2\nu_{su}t)] \\
& + [1 + \exp(-2\nu_{su}t)]^2 \log_2[1 + \exp(-2\nu_{su}t)] \\
& - [1 + \exp(-4\nu_{su}t)] \log_2[1 + \exp(-4\nu_{su}t)]\} .
\end{aligned}
\qquad (7.10)
$$

## 7.3   ESTIMATING THE ERROR FREQUENCY BEFORE CORRECTION

Applying the computed capacity formulas to actual genomes would of course imply that the kind of errors and their frequency are known. Unfortunately, this is not the case. Although many works were intended to estimate the mutation rate in genomes, none of them can directly be used for our purpose. Indeed, since the existence of genomic error-correcting codes is not recognized by current biology, the error frequencies were measured only on the basis of observed mutation rates hence, according to our point of view, *after* error correction. What we need is the error frequency *before* correction.

We now examine how the frequency of errors can be estimated. Due to the relevance of mutations to pathologies, the human genome has been especially studied in this respect: by Haldane, as early as 1935 [64], and recently by Kondrashov [52] and many others. They all agree on a figure of about $2 \times 10^{-8}$ error per nucleotide per generation in the average, the errors mainly consisting of substitutions. Interestingly, its reciprocal is 50 million generations, or about 1.5 billion years. This is the proper order of magnitude although this figure is somewhat too small: remember that the origin of life dates back to at least 3.5 billion years.

Although it concerns an error rate after correction, we may try to use this figure in order to estimate the error frequency before correction, but doing so needs some further assumptions. We were already led to the subsidiary hypothesis that the genomic code assumes a nested structure implying a very unequal protection of the nucleotides against errors, depending on their location. We were moreover led to conjecture, for making the genomic code compatible with sexual reproduction, that the outer layer of the nested system is left uncoded (see Sec. 9.4). Let $u$ denote the proportion of nucleotides which belong to this layer and are thus left uncoded. If the overall mutation rate per nucleotide per generation is denoted by $\mu$, we have $\mu = uv + (1 - u)\varepsilon$ where $v$ denotes as above the error frequency per nucleotide (expressed using a generation as time unit) and $\varepsilon$ denotes the average error rate after regeneration which is contributed to the error rate by the inner, encoded layers of the nested system. As an approximation we may reasonably assume that $\varepsilon$ can be ignored, which results in $\mu = uv$. The polymorphism of the human genome concerns about one thousandth of the nucleotides. Then, letting $\mu = 2 \times 10^{-8}$ and $u = 10^{-3}$ results in a frequency of $v = 2 \times 10^{-5}$ error per nucleotide per generation, a fairly plausible estimate in the absence of direct experimental data. That this estimate relies on data from the human genome is unlikely to restrict its generality, since the type and frequency of errors are presumably intrinsic properties of DNA in its terrestrial environment, regardless of the organism which hosts it. The time constant $1/v$ is then about 50,000 generations or about 1.5 million years. It is very large at the scale of a human life hence can well give the illusion that genomes are permanent objects, although it is short at the geological timescale.

The assumption of a constant frequency of errors is not necessary if we only look for an upper bound on the channel capacity. It suffices to use the smallest value it can assume to obtain an upper bound on the actual capacity. Similarly, as regards the type of errors it is possible to merely compute an upper bound on the DNA capacity, based on hypotheses which systematically minimize the impact of errors. The mildest kind of error is erasure, and an upper bound on the DNA capacity would result from assuming that all errors are of this type. Then, the true capacity of DNA would be less than that which concerns erasures and takes the double-strand structure into account, as given by Eq. (7.9) and illustrated in Fig. 3.11. This bound decreases exponentially down to 0 as time increases, and the actual DNA capacity is less.

## 7.4    PARADOXICALLY, A PERMANENT MEMORY IS EPHEMERAL

Regardless of the uncertainty about the type and frequency of the errors, the above computations show that the DNA capacity vanishes exponentially fast whatever assumptions are made. DNA is indeed an ephemeral memory, hence unable to conserve the hereditary information during geological times without the help of intrinsic error-correcting means used frequently enough to efficiently regenerate the genome.

This conclusion is by no means specific to DNA. Like any 'permanent' memory, DNA is paradoxically ephemeral because it faithfully keeps the incurring errors as well as its initial content. Errors accumulate up to eventually replacing the initial memory content. The above calculations show moreover that its information-theoretic capacity vanishes exponentially fast. The need for regenerating the memory content from time to time is actually shared by any 'permanent' memory used in the presence of errors. Stated in more adamant words, 'no memory is permanent.' The engineers who design computer memories know that. They have recourse to dynamic processes generally involving error-correcting codes to increase the actual permanence of the recorded data up to the needed value. There are moreover concerns as regards conservation of recorded data beyond a few decades. Compare with the few billion years of geological times!

CHAPTER 8

# A Toy Living World for Illustrating the Main Hypothesis

We now introduce a very simple model, referred to as 'toy living world,' where $n$-symbol sequences referred to as 'genomes' are subjected to random errors. The 'genomes' can be either arbitrary sequences or words of an error-correcting code. They are periodically replicated, after regeneration when an error-correcting code enables it. The model is as simple as to enable easy calculations of the lifetime of a 'genome' and the population size of the corresponding 'species.' If no genomic error-correcting code is used, no discrete species can exist beyond a certain genome length: a property of the true living world as fundamental as the existence of discrete species is lost, which shows the inadequacy of the uncoded model. It turns out on the contrary that the 'toy living world' shares basic features with the actual one in the case where a genomic error-correcting code is used, especially as regards the existence of discrete species. The toy living world just illustrates the main hypothesis. The following chapter will deal with the subsidiary hypothesis that the genomic code takes the form of a 'nested system.'

## 8.1   A SIMPLE MODEL

According to our *main hypothesis*, genomes act as error-correcting codes, and we have seen above (Ch. 7) that the vanishing capacity of DNA makes it absolutely necessary. We shall first illustrate this main hypothesis with a model as simple as to enable easy calculations, but hopefully realistic enough to mimic some basic features of the actual living world.

Assuming regeneration and replication to be jointly performed, an insight in the transmission of genetic information can be gained by analyzing a 'toy living world' in which 'genomes' consist of words of some binary error-correcting code of length $n$ (we throughout put 'genomes' between quotes to remind their fictitious character). We refer to the code as to the 'genomic code.' The 'genomes' are periodically regenerated and replicated in the presence of symbol substitutions occurring with a constant probability $p$, giving rise to 'genome' populations. All identical 'genomes' are said to belong to a same 'species' and their number at a given instant is referred to as the 'population' of this 'species.' For comparison, we also consider the case where no intrinsic genomic code exists hence where a 'genome' is an arbitrary $n$-bit sequence. Since it cannot be regenerated, only its replication is performed. In order to use a single word in both cases, we refer to regeneration followed with replication as 'duplication' in the case where an error-correcting code is used, the same word being used as a synonymous of 'replication' in the uncoded case. We denote by $\Delta t$ the time interval between successive duplications. For a constant symbol error frequency $\nu_{su}$, the probability of symbol error is $p = p_{su}(\Delta t)$, where $p_{su}(t)$ is given by Eq. (7.3).

In a first step, we do not consider any limit to the population growth, hence no natural selection. When a genomic error-correcting code is assumed to exist, a regeneration error results in a 'genome' which differs from the original one in at least $d$ symbols, the minimum Hamming distance of the genomic code. The occurrence of regeneration errors thus results in discrete 'species' successively appearing. Stemming from a single ancestral 'genome,' regeneration (decoding) errors generate an increasing number of new 'genomes' according to a branching process. In the uncoded case, on the contrary, 'genomes' are comparatively short-lived random sequences which do not constitute discrete 'species' beyond some critical length.

We first develop an analysis of this 'toy living world.' In a second step we introduce in it natural selection. A third step could consist of taking the subsidiary hypothesis (that the genomic error-correcting code takes the form of a nested system) into account. Doing so would introduce a hierarchical taxonomy and result in a model of the living world much more similar to the real one but much less simple. We shall deal with the subsidiary hypothesis more roughly, only qualitatively relying on the analysis of the simple toy living world based on the main hypothesis, in the next chapter.

## 8.2    COMPUTING STATISTICAL QUANTITIES RELEVANT TO THE TOY LIVING WORLD

We now analyze the toy living world exclusively based on the main hypothesis, assuming 'genomes' of a given length $n$, each symbol of which incurs an error independently of the others with a probability $p$ which depends only on the error frequency $\nu$ and the time interval $\Delta t$ since the last duplication. The 'genomes' either are not encoded or belong to some error-correcting code of minimum Hamming distance $d$. In the latter case, the error-correcting code enables regenerating the 'genomes.' Remember that 'duplication' is intended to mean either regeneration followed by replication, or replication alone in the uncoded case. It provides two copies of the possibly regenerated 'genome.'

We say that a 'genome' incurs a mutation when it differs from the original one. Let $P$ denote the probability of this event. In the absence of error correction, this probability is the complement to 1 of the probability that the 'genome' is received without error, hence $P = P_{unc} = 1 - (1 - p)^n$. Then $P_{unc}$ is fairly large for reasonable values of $p$ and $n$. Moreover, it is an increasing function of the 'genome' length $n$. When an error-correcting code is used, the probability of a mutation, i.e., of the regeneration process failing to recover the original 'genome,' $P = P_{cod}$, can on the contrary be made *arbitrarily small* through the choice of a long and efficient enough code. Moreover, $P_{cod}$ can be made the smaller, the larger the codeword length $n$. The precise value of $P_{cod}$ depends on the genomic code so no simple expression of it can be given. However, a plausible approximation of it is $Kp^{d/2}$ where $K$ is a constant which depends on the code. Insofar as the error-correcting code is efficient, $P_{cod}$ is much smaller than $P_{unc}$.

The following calculations are valid whether an error-correcting code is used or not which helps understanding the specific features brought by encoding. $P$ standing below for $P_{cod}$ or $P_{unc}$, depending on the case, the following results hold:

- the probability that a 'genome' is conserved (i.e., remains identical to itself) after $i$ successive duplications is $(1 - P)^i$;

- the probability that a 'genome' fails to be regenerated after exactly $i$ successive successful regenerations or errorless duplications is $P(1 - P)^i$. This is also the probability that a 'genome' lasts exactly $i \Delta t$, or that its 'lifetime' assumes this value, where $\Delta t$ is the time interval between duplications;

- the average lifetime $L(P)$ of a 'genome,' expressed using $\Delta t$ as time unit, to be referred to in the sequel as its *permanence*, is the expectation of its lifetime $i$ affected with its probability $P(1 - P)^i$, namely

$$L(P) = \sum_{i=1}^{\infty} i P(1 - P)^i = (1 - P)/P . \tag{8.1}$$

When an adequate error-correcting code is employed, and if $\Delta t$ is short enough, $P = P_{cod}$ is very small so $1/P_{cod}$ is a simple approximation to the 'genome' permanence. This permanence increases without limit if we use a long and efficient enough error-correcting code such that $P_{cod}$ approaches 0. This is possible provided the condition stated by the fundamental theorem of channel coding that the source entropy is less than the channel capacity is satisfied (see Sec. 5.3.1). Then the average lifetime of a 'genome' can assume values of the order of magnitude of geological times. If no error-correcting code is used, $P = P_{unc} = 1 - (1 - p)^n$ so the permanence $L(P_{unc})$ is approximated by $1/[1 - (1 - p)^n]$ or, for $p$ small enough (i.e., if $p \ll 1/n$), by $1/np$. The permanence of a 'genome' is then the shorter, the longer it is, at variance with the encoded case;

- the total number of 'genomes' which remain identical to the initial one after $i$ duplications, hence the population of the corresponding species at that time, is in the average $[2(1 - P)]^i$ since we assume that each duplication implies a replication which results in two identical 'genomes.' In the absence of any limiting factor, this number exponentially increases only if $P < 1/2$. If no error-correcting code is used this inequality is satisfied only for short enough 'genomes,' such that $P_{unc} < 1/2$, which implies the inequality $(1 - p)^n > 1/2$. Taking logarithms to the base 2 results in $n < -1/\log_2(1 - p)$. An approximation of $-\log_2(1 - p)$ for $p$ small enough is $p/\ln(2)$, so the existence of stable species if no encoding is performed is only possible for 'genome' lengths $n$ satisfying the inequality $n < \ln(2)/p \approx 0.69/p$. For larger values of $n$, the 'genomes' obtained would be diverse and constantly changing, without any definite species. If on the contrary an error-correcting code is used, $P_{cod}$ can be made much smaller than 1/2 and the persistence of a species is ensured by the exponential increase of the population of individuals which share the same 'genome.'

Before discussing some other properties of the toy living world and making it more realistic by taking into account natural selection, let us notice that the above calculations unambiguously show that the absence of coding is not compatible with the existence of discrete species with long genomes, at variance with the actual living world. The permanence $L(P_{\text{unc}})$ of a species in the absence of coding has been shown above to decrease proportionally to the inverse $1/n$ of the genome length. Moreover, the population of a species grows (hence its persistency is secured) only provided the following condition holds:

$$n < \ln(2)/p \,. \tag{8.2}$$

## 8.3   THE INITIAL MEMORY CONTENT IS PROGRESSIVELY FORGOTTEN

It is easy to show that any $n$-symbol word can result from errors accumulated during a large enough number of duplications, up to the point where the initial word is actually 'forgotten.' By 'any $n$-symbol word' we mean an arbitrary binary sequence of length $n$ in the absence of coding, or any word of the genomic code when an error-correcting code is used.

Let us first consider the uncoded case. We assume that a 'genome' is replicated $g$ times, and that each time it is replicated a given symbol has some probability $p$ to incur an error, i.e., to be changed into its binary complement. We assume that such an error is truly random, i.e., that $p$ differs from both 0 and 1: $0 < p < 1$. Moreover, we assume that the successive errors occur independently from each other. We wish to compute the probability $P_0$ that, after $g$ replications, a given symbol is the same as the original one. This occurs if and only if the number of errors it incurs within the $g$ successive generations is zero or even. The probability of $i$ errors within $g$ generations equals $\Pr(i) = \binom{g}{i} p^i (1-p)^{g-i}$, so $P_0$ is the sum of these probabilities for $i$ zero or even. Notice that $\Pr(i)$ can be obtained as the $i$-th term (starting with $i = 0$) of the development of $(x + y)^g$, letting $x = 1 - p$ and $y = p$. (Incidentally, we check that $\sum_{i=0}^{g} \Pr(i) = 1$, as expected.) The probability $P_0$ is the sum of the terms in the development of $(x + y)^g$ where the power of $y$ is zero or even. Now consider $(x - y)^g$. In its development, the terms where the power of $y$ is zero or even are the same as in the development of $(x + y)^g$, while those where the power of $y$ is odd have the opposite sign. Then $(x + y)^g + (x - y)^g$ gives twice $P_0$. Letting $x = 1 - p$ and $y = p$ results in:

$$P_0 = \frac{1}{2}[1 + (1 - 2p)^g] \,.$$

Since $0 < p < 1$ entails $|1 - 2p| < 1$, $(1 - 2p)^g$ tends to 0 as $g$ increases. $P_0$ thus becomes the closer to 1/2, the larger $g$. After a large enough number $g$ of replications, each binary symbol is thus random with a probability of about 1/2, independently of the others. It no longer gives any clue on the initial value of the symbol, which we may think of as *forgotten*. The resulting set of 'genomes' tends to a large number of purely random sequences. In sharp contrast, a species of the toy living world is a set of identical 'genomes.'

The case where an error-correcting code is used is less simple. The probability $P_{cod}$ of a regeneration error, for an efficient enough code and a short enough time interval between successive regenerations, is much smaller than the probability $P_{unc}$ of an error without such a code. $P_{cod}$ depends in a complex way on the parameters of the code. The regeneration errors are much less frequent than in the absence of an error-correcting code, but a regeneration error entails that at least $d$ symbols among the $n$ symbols of the codeword are in error ($d$ is the minimum Hamming distance of the code). If a large number of regeneration errors occur (which implies a very long time interval, say at the geological scale), a similar phenomenon occurs, i.e., an average of one symbol out of two becomes wrong and the 'genome' becomes again asymptotically unrelated to the original one. Let $e(i)$ denote the number of erroneous symbols after $i$ regeneration errors. If one more regeneration error occurs, this number becomes

$$e(i + 1) = e(i)(1 - d/n) + (d/n)[n - e(i)] = e(i)(1 - 2d/n) + d ,$$

assuming as an approximation that a regeneration error entails exactly $d$ erroneous symbols. This expression results from noticing that, when an error affects a binary symbol, it becomes wrong if it is correct but is corrected if it was already wrong. It relies moreover on the simplifying assumptions that exactly $d$ erroneous symbols result from a regeneration error, and that the successive error patterns which result from regeneration errors are mutually independent. Both are not exactly true but mere approximations. According to the above expression, $e(i)$ approaches $n/2$ regardless of the minimum distance $d$ of the code as $i$ tends to infinity (incidentally, this result is valid for $d = 1$, hence in the uncoded case, too). The use of a genomic error-correcting code makes however a big difference with the uncoded case since the result of a regeneration error is a codeword. Although the 'genome' eventually obtained results from a series of random choices, it is not itself a purely random sequence of symbols since it obeys the specific constraints of the code to which it belongs. Although each regeneration error results in a random jump to a genome at a Hamming distance at least $d$ apart when an error-correcting code is used, the overall speed of evolution is then much slower than in the uncoded case since regeneration errors are very seldom: it proceeds with larger steps which however occur very much less often.

Notice that in the toy living world discrete 'species' originate in regeneration errors just because the 'genome' is a word of an error-correcting code, apart from any selection. The average lifetime $L(P_{cod})$ varies proportionally to the inverse of the regeneration error probability $P_{cod}$. Hence, the less likely is a mutation, the larger the permanence of the mutated 'genome.' The genomic message which results from a mutation is made just as resilient to casual errors as the original one since both are codewords of the same genomic error-correcting code.

For the purpose of illustration, we'll consider in Sec. 8.5 a toy living world using a very simple code and show some results of its simulation.

## 8.4    INTRODUCING NATURAL SELECTION IN THE TOY LIVING WORLD

Of course, the toy living world analyzed above is oversimplified. It could maybe model the evolution of a bacterial population in very favorable conditions and in the absence of any 'horizontal transfer' of genetic material. The individuals are assumed to be eternal and no kind of natural selection is considered. Even for potentially eternal individuals, the population increase in the real world would become rapidly slower than exponential as a mere result of the finiteness of available resources, according to the usual scheme of Darwinian selection. We now assume that, analogously to the actual living world, some specific phenotype hosts each 'genome' and is subject to environmental factors more or less favorable to its survival so as to hopefully make the toy living world a bit more similar to the actual one. This case can be dealt with by introducing a probability of death, $P'$, besides the probability $P$ of a mutation. 'Death' means here that the 'genome' is not conserved for another reason than the occurrence of a mutation due to channel errors, for instance its recombination or the destruction of the phenotype which hosts it.

The above computation of the probability that the 'genome' is not conserved is easily extended. It suffices to notice that the survival of the 'genome' as regards its death (in the above meaning) and the absence of a mutation affecting it are mutually independent events, so the probability of the 'genome' conservation is the product of the probabilities of these two events. These probabilities are $1 - P'$ and $1 - P$, respectively, so the 'genome' is conserved with probability $(1 - P)(1 - P') = 1 - P - P' + PP'$. The probability that it is not conserved is thus $P + P' - PP'$. All the results computed in Sec. 8.2 remain valid in the case where natural selection occurs provided the probability $P$ is replaced by $P + P' - PP'$. For instance, the permanence (average lifetime) of the 'genome' becomes

$$L(P, P') = \frac{1 - P - P' + PP'}{P + P' - PP'}, \tag{8.3}$$

closely approximated by $1/(P + P' - PP')$ if $P$ and $P'$ are small enough. One easily checks that $L(P) = L(P, 0) > L(P, P')$ for any $P' > 0$, as expected.

The difficulty of properly evaluating $P'$ in the real world, which obviously depends on very many factors, limits however the practical usefulness of this result. The death probability $P'$ then mainly depends on how well the phenotype corresponding to the considered 'genome' fits its environment. In the absence of error correction, it is only the phenotype fitness which would ultimately cause some species to survive rather than another one. Since mainstream biology ignores error-correcting codes, it is the sole mechanism it contemplates for explaining the discreteness of species, but there is no reason why natural selection should carve well separated and very sparse 'genomes' out of the mass of possible nucleotide sequences. If, on the contrary, a genomic error-correcting code is used, a new 'genome' results from a regeneration error. The nascent species originating in it becomes the target of natural selection, but its discreteness is an intrinsic preexisting property of the genomic code, not the sole consequence of natural selection.

Our main hypothesis thus suffices to account for the existence and evolution of 'species,' even in the absence of any natural selection. In sharp contrast, natural selection is thought of according to the neo-Darwinian orthodoxy as the sole reason why discrete species exist, as discussed, e.g., by Mark Ridley [69]. Neo-Darwinians assert that natural selection results in sharply defined species rather than poorly differentiated living things. This statement is, however, not supported by plausible arguments. There are cases where natural selection operates only within the limits of some hereditary invariant. For instance, the limbs of many vertebrates have 5-digit extremities. This fact cannot result from an adaptation but from the permanence of a hereditary information, since the functions achieved widely differ from a species to another: think of the hand and the foot of a man, the foot of a lizard or a mole, the fin of a dolphin or a whale, the wing of a bat, etc., fitted to as diverse functions as prehension, walking, running, digging, swimming, flying, etc. This adaptation results in considerable variations in the form, size and mutual connection of bones, tendons, muscles, skin …*within* the 5-digit pattern, which appears as a hereditary invariant.

## 8.5    EXAMPLE OF A TOY LIVING WORLD USING A VERY SIMPLE CODE

We present here a very simple example of a 'toy living world.' Since the regeneration performance depends on the total number of erroneous symbols in a word, we may assume a very short 'genome' and a very high error rate. We ran simulations assuming a binary 'genome' of length $n = 7$ and a symbol error rate of $p = 0.1$. (In a realistic situation, a given error rate would be obtained by properly adjusting the time interval $\Delta t$ between successive regenerations, for a given error frequency.) We used either no error-correcting code, or the very simple (7,4) Hamming code which can correct all single errors (see Ch. 6). The probability of a regeneration error is then $P_{\mathrm{cod}} = 1 - (1 - p)^7 - 7p(1 - p)^6$ which approximately equals 0.16 for $p = 0.1$. We assume that no other events than replications-regenerations occur. Especially, we assume that no natural selection limits the number of replicated 'genomes.' Despite the extreme simplicity of this model and the lack of natural selection, i.e., the factor which is believed the most important for shaping the living world, what we obtain mimics it rather well. Moreover, introducing natural selection (Sec. 8.4) and taking account of the subsidiary hypothesis (see Ch. 9 below) would refine the model and make it closer to the real living world, although of course still oversimplified.

A drawback of the choice of the very short (7,4) Hamming code is that the average number of erroneous bits in a received word is small, hence subject to large statistical fluctuations. The stability of species in the toy living world is thus much less than that of a model more realistically involving much longer genomes. Apart this important exception, we hope that analyzing the toy living world sheds some light on the properties which may be expected from the main hypothesis that genomic error-correcting codes exist.

The Hamming (7,4) code is the first invented nontrivial code (in the late 1940s). It uses the binary alphabet ($q = 2$). The length of its words is $n = 7$. The constraints which tie together its bits

$c_1, c_2, \ldots, c_7$ are

$$c_1 \oplus c_3 \oplus c_4 \oplus c_5 = 0$$

$$c_1 \oplus c_2 \oplus c_3 \oplus c_6 = 0$$

and

$$c_2 \oplus c_3 \oplus c_4 \oplus c_7 = 0$$

where $\oplus$ denotes addition modulo 2. It comprises only $2^4 = 16$ words which are listed in Tab. 8.1 below and numbered from 0 to 15:

| Table 8.1: The 16 words of the (7,4) Hamming code | | | |
|---|---|---|---|
| 0: | 0 0 0 0 0 0 0 | 8: | 0 0 0 1 1 0 1 |
| 1: | 1 0 0 0 1 1 0 | 9: | 1 0 0 1 0 1 1 |
| 2: | 0 1 0 0 0 1 1 | 10: | 0 1 0 1 1 1 0 |
| 3: | 1 1 0 0 1 0 1 | 11: | 1 1 0 1 0 0 0 |
| 4: | 0 0 1 0 1 1 1 | 12: | 0 0 1 1 0 1 0 |
| 5: | 1 0 1 0 0 0 1 | 13: | 1 0 1 1 1 0 0 |
| 6: | 0 1 1 0 1 0 0 | 14: | 0 1 1 1 0 0 1 |
| 7: | 1 1 1 0 0 1 0 | 15: | 1 1 1 1 1 1 1 |

Notice that the number made of the first four bits of each word is the binary representation of the word number (with the less significant bits on the left). This is referred to as the 'systematic form' of the code (see Sec. 6.4.2 in Ch. 6).

This code has been designed according to a precise algebraic structure, but to understand how it works, it suffices to notice that its minimum distance is $d = 3$, which means that it can definitely correct any substitution of a single binary symbol for its binary complement, or the erasure of any 2 binary symbols (see Ch. 6).

As an illustration of a hereditary process where a 'genome' is successively replicated several times starting from an ancestral one, and of the introduction of an error-correcting code in this process, we consider a 'toy living world' whose 'genome' is binary and of very short length, namely 7. We assume that the error rate is very high, namely $p = 0.1$, with the errors drawn at random independently of each other. Then the probability that an error pattern of weight $e$ occurs in a word of length 7 is $P(e) = \binom{7}{e} p^e (1 - p)^{7-e}$. The numerical values of $P(e)$ are given in the following table.

| Table 8.2: Probability $P$ of an error pattern of weight $e$ | | | | | | | |
|---|---|---|---|---|---|---|---|
| $e$ | 0 | 1 | 2 | 3 | 4 | 5 | 6 | 7 |
| $P$ | 0.4782969 | 0.3720087 | 0.1240029 | 0.0229635 | 0.0025515 | 0.0001701 | 0.0000063 | 0.0000001 |

Simulating this process on a calculator, we obtained the following sequence of 63 error patterns of length 7 (an erroneous symbol is indicated by 1):

0000000 0000000 0000000 0101000 0000000 0000000 0000001 0100000
0000101 0000001 0000000 0010000 0000011 0000000 0100000 0010000
0000000 0000000 0000010 0000001 0000000 0000000 0000000 0000000
0000000 0000000 0000010 0000010 0000000 0010000 0000000 0001000
0000000 0100000 0001000 0000000 0000000 0000010 0001111 0000010
1000000 0100000 0000000 0000100 1100000 0000000 1000000 0000100
0000000 0000010 0000000 0000100 0000000 0000000 1000000 0000000
0000010 0000000 0001001 0000001 0001000 0000000 0000001.

Assuming the ancestral 'genome' to be 1111111, we considered the evolution which results from successive replications of this 'genome' in the presence of mutations due to the above error patterns, either without error-correction or when the (7,4) Hamming code is employed (1111111 is a word of this code). This code has three as minimum distance. Therefore, it corrects all error patterns with a single erroneous symbol. However, all error patterns involving two or more erroneous symbols result after decoding in an error pattern with at least three erroneous symbols. The 'genealogical trees' obtained in both cases are represented in the following two figures.

A sample of the simulated evolution of the population of 'genomes' when no error-correcting code is used is given in Fig. 8.1. The population of 'genomes' contains more and more different individuals as the number of successive replications increases. They are short-lived genomes and no distinct species can be observed. The initial 'genome' rapidly ceases to be a majority in this population which, after a few replications, looks like a set of random words. When the (7,4) Hamming error-correcting code is used, the initial 'genome' remains present during a larger number of regenerations-replications. Regeneration errors give rise to 'genomes' differing from the initial one by at least 3 symbols, which may be interpreted as other 'species.' Moreover, they exhibit the same permanence as the initial 'genome.'

## 8.6    EVOLUTION IN THE TOY LIVING WORLD; PHYLETIC GRAPHS

Notice that Figs. 8.1 and 8.2 represent genealogical trees, the elements of which are individuals. We consider as a 'species' the set of identical 'genomes' (the model does not take into account the slight individual differences which exist in the real world). One can deduce *phyletic graphs* from the genealogical trees by merging into a single branch all branches which correspond to a same 'genome.' Doing so for the genealogical tree of the uncoded case, represented in Fig. 8.1, results in a rather messy graph. The phyletic graph in the encoded case, deduced from the genealogical tree of Fig. 8.2, is more interesting. It is drawn in Fig. 8.3. New species originate in regeneration errors, represented as dots in the figure. One notices that some branches converge, so this graph is not a tree. The code of the example contains however very few words so there is a high probability that distinct regeneration errors result in the same 'genome,' making such convergences likely events.

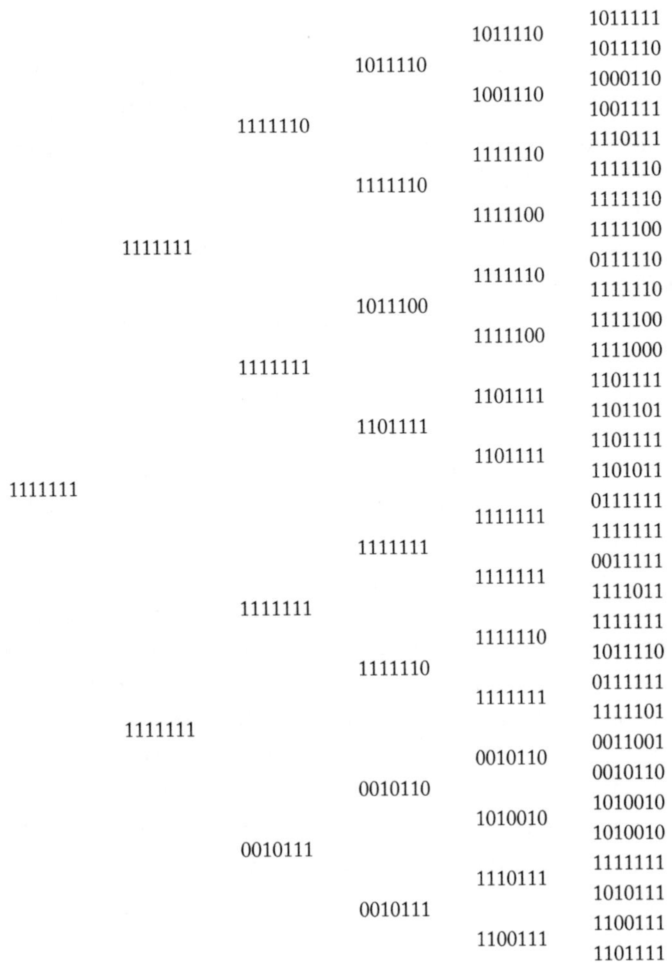

```
                                                                      1011111
                                                          1011110     1011110
                                              1011110                 1000110
                                                          1001110     1001111
                                  1111110                             1110111
                                                          1111110     1111110
                                              1111110                 1111110
                                                          1111100     1111100
                      1111111                                         0111110
                                                          1111110     1111110
                                              1011100                 1111100
                                  1111111                 1111100     1111000
                                                          1101111     1101111
                                              1101111                 1101101
                                                          1101111     1101111
          1111111                                                     1101011
                                                          1111111     0111111
                                              1111111                 1111111
                                                          1111111     0011111
                                  1111111                             1111011
                                                          1111110     1111111
                                              1111110                 1011110
                                                          1111111     0111111
          1111111                                                     1111101
                                                          0010110     0011001
                                              0010110                 0010110
                                                          1010010     1010010
                                  0010111                             1010010
                                                          1110111     1111111
                                              0010111                 1010111
                                                          1100111     1100111
                                                                      1101111
```

**Figure 8.1:** Genealogical tree of the toy living world, without error correction.

This probability is much lower for codes having more numerous words close to each other in terms of the Hamming metric, except maybe for the very redundant codes in the innermost layers of the nested system (see Ch. 9). This case excepted, it may be expected that branch convergences are highly unlikely so all phyletic graphs met in practice assume the shape of a tree: branch convergences are highly improbable, although not strictly impossible.

The graph in Fig. 8.3 illustrates the relationship between error-correction coding and evolution for the toy living world: new species originate in regeneration errors. If we ignore the branch convergences (represented by arrows) which occur due to the choice of a very short and simple code

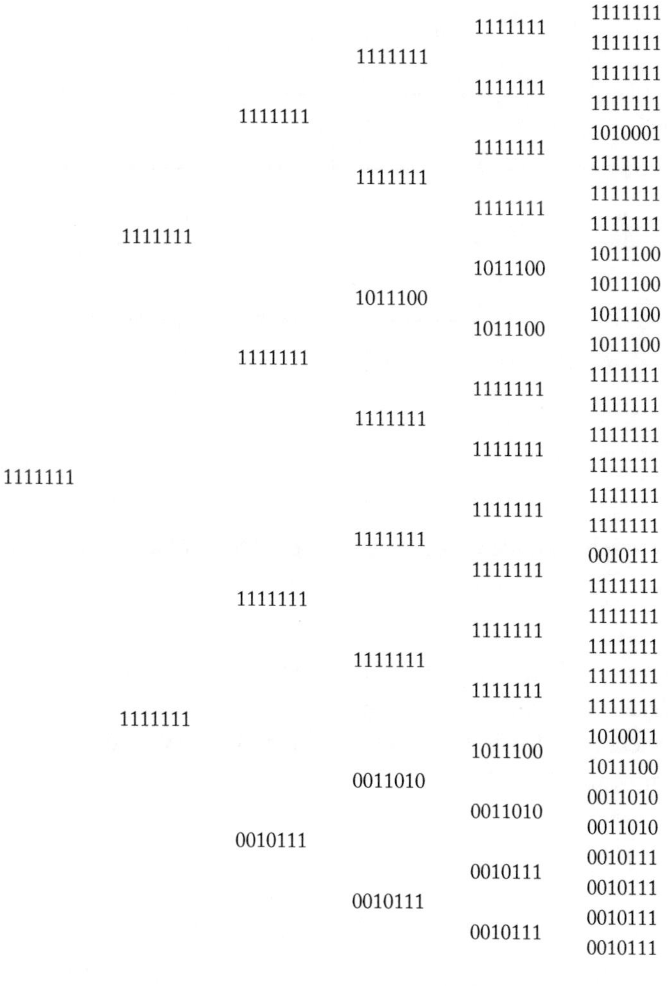

| | | | | 1111111 | 1111111 |
| | | | 1111111 | | 1111111 |
| | | | | 1111111 | 1111111 |
| | | | | | 1111111 |
| | | 1111111 | | 1111111 | 1111111 |
| | | | | | 1010001 |
| | | | 1111111 | 1111111 | 1111111 |
| | | | | | 1111111 |
| | | | | 1111111 | 1111111 |
| | 1111111 | | | | 1111111 |
| | | | | 1011100 | 1011100 |
| | | | 1011100 | | 1011100 |
| | | | | 1011100 | 1011100 |
| | | 1111111 | | | 1011100 |
| | | | | 1111111 | 1111111 |
| | | | 1111111 | | 1111111 |
| | | | | 1111111 | 1111111 |
| 1111111 | | | | | 1111111 |
| | | | | 1111111 | 1111111 |
| | | | 1111111 | | 1111111 |
| | | | | 1111111 | 0010111 |
| | | 1111111 | | | 1111111 |
| | | | | 1111111 | 1111111 |
| | | | 1111111 | | 1111111 |
| | | | | 1111111 | 1111111 |
| | 1111111 | | | | 1111111 |
| | | | | 1011100 | 1010011 |
| | | | 0011010 | | 1011100 |
| | | | | 0011010 | 0011010 |
| | | 0010111 | | | 0011010 |
| | | | | 0010111 | 0010111 |
| | | | 0010111 | | 0010111 |
| | | | | 0010111 | 0010111 |
| | | | | | 0010111 |

Figure 8.2: Genealogical tree of the toy living world using the (7,4) Hamming code.

but would be very unfrequent for a more realistic code choice, this graph depicts the evolution as a radiative process starting from the ancestral 'genome.' Due to a regeneration error, a chance event which moreover occurs at a random instant, the 'genome' of a new species is chosen among the codewords. Once a new species has been created, it potentially lasts indefinitely (just as the ancestral one) since its extinction due to mutations would imply the occurrence of simultaneous regeneration errors in all the individuals of the species. Its probability is thus $P_{cod}^j$, where $j$ denotes the number of individuals in the species and $P_{cod}$ is the probability of a regeneration error. Since replication is assumed to follow each regeneration, the smallest possible value of $j$ is 2. Moreover, $j$ grows expo-

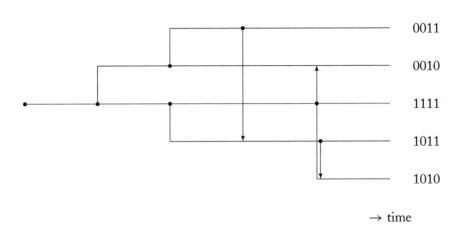

Figure 8.3: Phyletic graph associated with the genealogical tree of Fig. 8.2. Each 'species' is labeled with the four leftmost bits of its 'genome,' i.e., its information bits.

nentially as time passes, making the probability of extinction of a species due to mutations, $P_{\text{cod}}^{j}$, a miraculous event. Indeed, the extinction of a species can only occur when natural selection operates on the associated phenotype; then, it depends on the probability $P'$ of Sec. 8.4 and results from a differential lack of fitness.

CHAPTER 9

# Subsidiary Hypothesis, Nested System

This short chapter is entirely devoted to the concept of nested system basic to the subsidiary hypothesis. Its consequences on the shape of the living world are very important.

## 9.1 DESCRIPTION OF A NESTED SYSTEM

The main hypothesis that a genomic code exists should be supplemented by the *subsidiary hypothesis* that it actually combines several codes in a *nested system*. Indeed, we saw in Ch. 7 that the capacity of DNA is a rapidly vanishing function of time. Would an error-correcting code equally protect all symbols, then an information would be the less faithfully conserved, the older it is. It is the contrary which is observed in the living world: some of the most faithfully conserved parts of the genome, especially the *HOX* genes, are among the oldest. The subsidiary hypothesis is intended to account for this fact. It assumes that the genomic error-correcting code actually consists of a combination of codes which were successively laid down during the geological ages. Every time a new coding layer has been appended, the information inside it has acquired a much smaller probability of regeneration error. Francis Crick's 'frozen accident' finds here a substantial content.

Prior to examine natural nested systems, we give a detailed example of how an engineer can design a nested system. We assume that we use binary block codes in systematic form[1] as its components. The $j$-th code is denoted by $C_j(n_j, k_j)$, where $n_j$ is its length and $k_j$ its 'dimension,' i.e., the length of the message it encodes. We moreover assume that the message $M_j$ it encodes explicitly appears as the leftmost $k_j$ bits of the word. We graphically depict a word of this code as

$$\boxed{\quad M_j \; \| \; R_j(M_j) \quad}$$

where $R_j(M_j)$ denotes the set of $n_j - k_j$ redundancy bits which the encoder computes in terms of the $k_j$ bits of the message $M_j$; the double vertical bar separates the $k_j$ information bits of this code from its $n_j - k_j$ redundancy bits. Using this representation, the following figure describes a four-layer nested system where the fourth layer is left uncoded.

A word belonging to the four-layer nested system is represented at the bottom of Fig. 9.1. Its construction involved four steps. In the first one (top), the $i_1$-bit information message $I_1$ is encoded according to a first code $C_1(n_1, k_1)$ with $k_1 = i_1$, i.e., the set of $n_1 - k_1$ redundancy bits denoted by $R_1(I_1)$ is computed in terms of $I_1$ and appended to it. In step 2, the $i_2$-bit information message $I_2$

---

[1] Remember that in a code in systematic form the information message explicitly appears at given locations in the codeword.

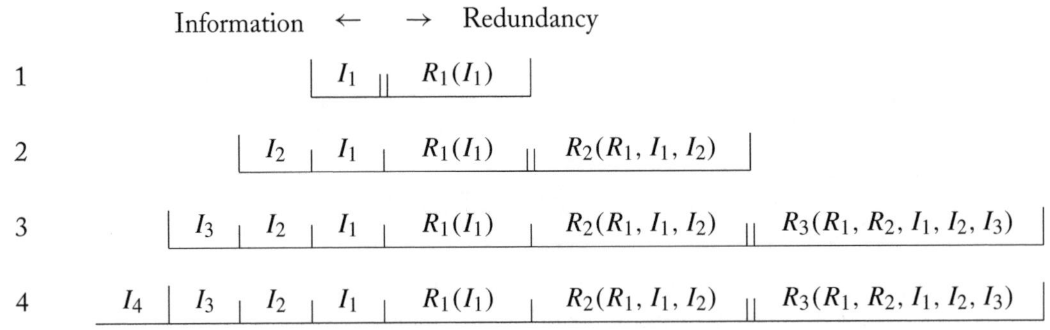

Information   ←   →   Redundancy

1   $\boxed{I_1 \parallel R_1(I_1)}$

2   $\boxed{I_2 \mid I_1 \mid R_1(I_1) \parallel R_2(R_1, I_1, I_2)}$

3   $\boxed{I_3 \mid I_2 \mid I_1 \mid R_1(I_1) \mid R_2(R_1, I_1, I_2) \parallel R_3(R_1, R_2, I_1, I_2, I_3)}$

4   $\boxed{I_4 \mid I_3 \mid I_2 \mid I_1 \mid R_1(I_1) \mid R_2(R_1, I_1, I_2) \parallel R_3(R_1, R_2, I_1, I_2, I_3)}$

**Figure 9.1:** A four-layer nested system. The mentions 'information' and 'redundancy' at the top refer to the overall redundancy of the nested system.

is appended to the codeword resulting from the encoding performed in step 1. The $(i_2 + n_1)$-bit message thus obtained is encoded by the code $C_2(n_2, k_2)$ with $k_2 = i_2 + n_1$, i.e., the set of $n_2 - k_2$ redundancy bits denoted by $R_2(R_1, I_1, I_2)$ is computed and appended. Similarly, in step 3 the $i_3$-bit information message is appended to the result of the encoding performed in step 2, resulting in an $(i_3 + n_2)$-bit message which is again encoded by the code $C_3(n_3, k_3)$ with $k_3 = i_3 + n_2$, i.e., the set of $n_3 - k_3$ redundancy bits denoted by $R_3(R_1, R_2, I_1, I_2, I_3)$ is computed and appended. The last step (step (4)) simply consists of appending the information message $I_4$ to the word resulting from the previous encoding since we assume that the fourth layer remains uncoded.

A simpler and more intuitive illustration uses the fortress metaphor: a code is represented as a wall which protects what is inside it against outside attackers. Several concentric walls have been successively built to enclose information, so the content of the oldest, most central, wall is much better protected than the more recent and peripheral information (see Fig. 9.2). Notice that a very efficient protection of the most central information does not demand very efficient individual codes: the multiplicity of enclosing walls is much safer than each of them separately.

None of the assumptions which we made to illustrate the nested system in Fig. 9.1 is mandatory. The alphabet needs not be binary. Its size can, moreover, be different from a coding layer to another one: a newly performed encoding may well deal with some extension of the information message, i.e., consider blocks of $m$ successive $q$-ary symbols as symbols of an alphabet of size $q^m$. The codes do not need to be in systematic form and, if they are, the location of the information bits is arbitrary. The assumption that the successive codes are of the block-code type may also be relaxed (convolutional codes are indeed likely candidates as being in a sense more 'natural' than block codes). In the case of convolutional codes the length of the codewords is no longer precisely defined, which moreover shows that even the condition that the dimension of each code is larger than the codeword length of the previous coding layer has not to be strictly satisfied. It suffices that the encoding performed at the $j$-th step brings some further protection to the message which has

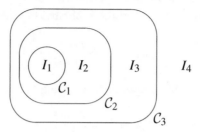

**Figure 9.2:** The same nested system as Fig. 9.1 represented according to the fortress metaphor. A code $\mathcal{C}_j$, $j = 1, 2, 3$, is represented as a closed wall which protects what is inside it. $I_1$, $I_2$, $I_3$, and $I_4$ are successive information messages. $I_1$ is protected by 3 codes, $I_2$ by 2 codes, $I_3$ by a single code, and $I_4$ is left uncoded.

already been encoded at the previous steps. Moreover, the codes actually used by nature do not result from an engineering design but likely belong to the broader class of 'soft codes', to be introduced in Ch. 10.

## 9.2 RATE AND LENGTH OF COMPONENT CODES

In order to get an insight on how the rate and length of the component codes vary when their number is large, we may assume that the length of the information message appended at each step of the construction of the nested system is constant, say $i$. We denote by $\rho_j = k_j/n_j$ the rate of the $j$-th component code. Its reciprocal $\lambda_j = 1/\rho_j = n_j/k_j$ will be referred to as the expansion factor of this code. The cumulated expansion factor for the first $j$ steps is defined as

$$\Lambda_j = n_j/ji$$

since then the cumulated information message has length $ji$. Clearly, the cumulated expansion factor at step $j$ obeys the recursion equality

$$j\Lambda_j = \lambda_j[(j-1)\Lambda_{j-1} + 1].$$

We may first assume, for instance, that $\lambda_j = \lambda$, a constant. Then we find that

$$\Lambda_j = (\lambda + \lambda^2 + \ldots + \lambda^j)/j,$$

or, in more compact form:

$$\Lambda_j = \frac{\lambda(\lambda^j - 1)}{j(\lambda - 1)},$$

which shows that the length $ji\Lambda_j$ of the $j$-th component code then grows exponentially in terms of the step number $j$.

It is, however, possible to choose the expansion factors $\{\lambda_j\}$ so as to limit the growth of the overall length. For instance, $\Lambda_j = \Lambda$, a constant, is obtained by choosing

$$\lambda_j = \frac{j\Lambda}{(j-1)\Lambda + 1}$$

which results in the rate of the $j$-th component code being

$$\rho_j = 1 - \frac{\Lambda - 1}{j\Lambda}.$$

Then the overall length of the system at step $j$ is $ji\Lambda$. If $\Lambda = 2$, for instance, the rate of the $j$-th component code is $\rho_j = (2j - 1)/2j$.

## 9.3   DISTANCES IN THE NESTED SYSTEM

Notice that the walls of the fortress of Fig. 9.2 are purely metaphoric. The presence of a wall just means that the elements enclosed in it are codewords separated by some minimum Hamming distance. Figure 9.3 attempts to visualize this distance distribution in the case of a two-layer nested system. Representing the distribution of points in a high-dimensional space is not easy and can only result in a very rough picture. First, the codewords are represented as points in the plane, according to some kind of projection, although they actually belong to an $n$-dimensional space. Second, the Hamming distance in the $n$-dimensional space is represented as the Euclidean distance in the plane of the figure. The restriction to a two-layer nested system aims at keeping the figure simple enough but a more realistic picture involving many nested codes would be extremely complicated.

We represented in Fig. 9.3 the first (older) code $C_1$ as a 'constellation' of points (big dots of the figure). At the second encoding layer, using the second code $C_2$ results in putting clusters of points centered on all the points of the initial constellation. It has been found convenient for drawing Fig. 9.3 to take each of these clusters exactly similar to the 'constellation' of points representing the codewords of $C_1$. Pursuing this process would result in a fractal structure. Although the actual structure of the nested system is probably much less simple, it is interesting to notice that the known correlation properties of the DNA sequences are compatible with a fractal structure.

The experimental analysis of DNA sequences has shown indeed they exhibit long-range dependence. Their power spectral density has been found to behave as $1/f^\beta$, asymptotically for small $f$, where $f$ denotes the frequency and $\beta$ is a constant which depends on the species: roughly speaking, $\beta$ is the smaller, the higher the species is in the scale of evolution; it is very close to 1 for bacteria and significantly less for animals and plants [75]. In another study of DNA sequences, the quaternary alphabet of nucleic bases $\{A, T, G, C\}$ was first restricted the to the binary one $\{R, Y\}$ by retaining only their chemical structure, purine or pyrimidine. An appropriate wavelet transform was used to cancel the trend and its first derivative. Then the autocorrelation function of the binary string thus obtained has been shown to decrease according to a power law [1], which implies long-range dependence (at variance with, e.g., Markovian processes which exhibit an exponential decrease).

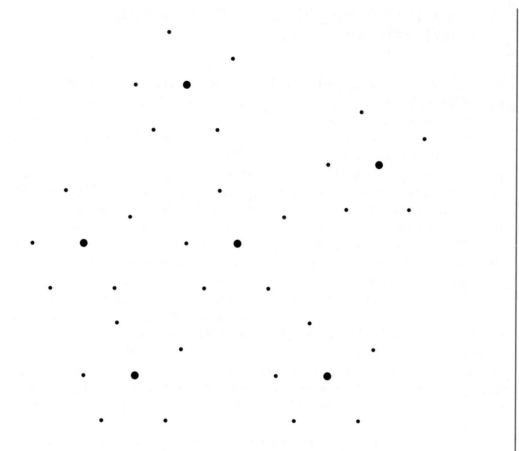

**Figure 9.3:**   Sketch of a two-component nested system in the space of sequences. The big dots represent the codewords of the first (older) component code $\mathcal{C}_1$, and all the dots represent words of the nested system, i.e., combining the first and the second component codes. The oldest information is borne by any of the points in the cluster which surrounds each of the big dots.

The $1/f^{\beta}$ behavior of the spectrum and the long-range dependence exhibited by the autocorrelation function of the DNA sequence are of course compatible with each other. Moreover, they are both compatible with a fractal structure, meaning that the DNA sequence may in some sense be self similar.

## 9.4    SOME CONSEQUENCES OF THE SUBSIDIARY HYPOTHESIS

We conjecture that individual genomic differences within a species are restricted to the peripheral layer of the nested system, assumed to be left uncoded as in Fig. 9.1. (Assuming that the outer layer of the nested system is left uncoded enabled in Sec. 7.3 estimating the nucleotide error rate in terms of measurements of the mutation rate.) Genomic encoding becomes compatible with sexual reproduction since then the genetic shuffling that crossing-over operates only affects the peripheral (actually uncoded) layer. Sex then promotes fast genetic changes in a population, an evolutive advantage against fast evolving pathogens or parasites, but confined in the peripheral layer of the nested system while the information of inner layers remains unaffected by symbol errors due to the genome error-correcting ability, except for very unfrequent regeneration errors.

The nested structure implies a very unequal protection of the nucleotides against errors, depending on their location. This makes the genome appear as very inhomogeneous as regards the susceptibility of its nucleotides to errors. The very unequal distribution of error rates depending on the location of the nucleotides implied by the subsidiary hypothesis matches the observed reality (see, e.g., [64, 52]). Interestingly, extremely well-conserved sequences of nucleotides are not only found in genes properly said (e.g., the *HOX* genes) but also in portions of the DNA which do not direct the synthesis of proteins (referred to as 'noncoding DNA' in the biological literature) [56]. Assuming that genomic error-correcting codes exist, we have no reason to restrict our attention to 'coding DNA.' On the contrary, any symbol in a codeword both contributes to, and benefits from, the error-correcting ability of the code. Needless to say, the idea of 'junk DNA' is entirely foreign to our hypotheses.

Analyzing the toy living world based on the use of a single code assumed to equally protect its symbols, as we did in Ch. 8, is a simple matter. A similar analysis of a model where the code is a nested system as just described would on the contrary be very complicated. Moreover, it would need several parameters like the number of component codes, their sizes and minimum distances which *a priori* depend on the species and are difficult to estimate. Even a toy living world limited to a two-component nested system would be difficult to analyze. We prefer to qualitatively describe the features of a world where the genomic code consists of a nested system.

Let us consider separately the component codes of the nested system. To begin with, consider a long and very redundant code as found in the deeper layers (the words 'deep' and 'peripheral' refer to the fortress metaphor). We may assume that it is good as a result of natural selection, 'good' meaning that its minimum distance is large, say equal to the Gilbert-Varshamov bound (see Eqs. (6.26) and (6.27) in Sec. 6.5.3). Being highly redundant, this code has comparatively few words separated from each other by large Hamming distances. In sharp contrast, we were already led to assume that the peripheral layer of the nested system is left uncoded. Then errors frequently occur but result in a genome very close to the original one. Between the deepest and the peripheral layers, the codes have less many codewords, the deeper the layer. At the same time, the codewords are the more separated in terms of the Hamming metrics, the code to which they belong lies in a deeper layer. The biological consequences of this nested structure will be more fully developed in Ch. 11.

CHAPTER 10

# Soft Codes

## 10.1 INTRODUCING CODES DEFINED BY A SET OF CONSTRAINTS

In Ch. 6, we defined an error-correcting code $\mathcal{C}(n, k)$ as a subset among the set $S_n$ of $n$-symbol sequences over some finite alphabet of size $q$. The number of symbols $k$ of the information message that each codeword represents is referred to as the number of dimensions of the code or, more briefly, as its dimension. The code being necessarily redundant implies the strict inequality $k < n$. There are basically two ways for defining such a code. One way is to give the list of its words; random coding, as used by Shannon and his followers in order to prove the fundamental theorem of channel coding, is a typical example of it. The trouble with a mere list of codewords is that the number of its elements exponentially grows in terms of the codeword length $n$: assuming a constant information rate $R = k/n$, $0 < R < 1$, the number of codewords is $q^k = q^{Rn} = (q^R)^n$, which makes the decoding process of such a code of prohibitive complexity when $n$ is large enough for the code to be useful. Each received word should indeed be compared with *all* the words of the code! Another way of defining an error-correcting code consists of stating some specific set of intrinsic constraints which its codewords should satisfy. All the sequences of length $n$ which satisfy this set of constraints, and only these sequences, belong to the code. The codewords can, in principle, be listed once this set of constraints is given, but the set of constraints is a much more parsimonious description of the code than the full list of its words.

The recourse to intrinsic constraints for defining an error-correcting code is an absolute necessity for making it tractable. Choosing a set of constraints without introducing a bias detrimental to the code performance is however a tough problem. It has even been thought of as unsolvable. A long-lived popular belief among communication engineers was expressed by the 'folk theorem': 'All codes are good, except those we can think of.' Its fallacy was only revealed with the invention of turbocodes. Presenting these codes in 1993, Berrou et al. reported unprecedented performance [21]. They showed that error-correcting codes based on a set of deterministic intrinsic constraints, but which mimic random coding as regards the distribution of distances between the codewords, exhibit a performance close to the theoretical limit, i.e., the channel capacity. Designing good codes is thus practically possible, although the contrary has long been believed true.

Another important advantage of defining an error-correcting code in terms of a set of constraints (as opposed to a list) is that these constraints are expressed as relations between the symbols of the codewords. Then the decoding procedure (i.e., aiming at recovering the original codeword from its received, possibly erroneous, version) does not need to deal with $n$-symbol sequences. Instead, decisions can be separately taken about each symbol and, moreover, provisional decisions on symbols can be used in the process of making new decisions, according to an iterative procedure. These possibilities are fully exploited in turbocodes, the decoding of which involves a kind of feedback made possible by symbol-by-symbol decoding (see Sec. 6.7.2).

## 10.2   GENOMIC ERROR-CORRECTING CODES AS 'SOFT CODES'

### 10.2.1   DEFINING SOFT CODES

Defining an error-correcting code in terms of a set of intrinsic constraints is an unavoidable necessity in engineering. The genomic error-correcting codes assumed to exist according to our main hypothesis do not escape it. However, this does not mean that natural codes must closely resemble those designed by engineers. The constraints considered by engineers are defined by deterministic *mathematical equalities* because they are thus perfectly defined and easily implemented using physical devices[1]. The implementation means of nature and humans widely differ, making the ease of engineering implementation irrelevant to nature. We thus propose to extend the kind of constraints which define an error-correcting code to inequalities, forbidding rules and constraints expressed in probabilistic terms, as well as physical-chemical constraints, besides mathematical equalities. Steric constraints due to the position and dimensions of molecules in the three-dimensional physical space are especially important. Still other constraints are linguistic. Genomes are assumed to direct the construction and maintenance of living things. This cannot be done without a language, which implies a number of constraints of morphological, syntactic and semantic character. According to this point of view, any constraint to which genomes or parts of genomes are subjected can be exploited to the benefit of error correction, regardless of its nature. We thus call for a wide extension of the concept of error-correcting code, to be referred to as *soft code*.

Even though we think that natural genomic error-correcting codes are not likely to closely resemble those designed by human engineers, the structure of eukaryotic genes with exons (used in the synthesis of a protein) and introns (having no known function) strikingly reminds that of block codes in systematic form which are currently used in engineering, where redundancy symbols only employed for correction purpose are appended to the information symbols. Are exons made of information symbols and introns of redundancy symbols? This question is examined in Sec. 10.5 below.

### 10.2.2   IDENTIFYING THE ALPHABETS

In engineering problems, the alphabet is often *a priori* given and endowed with some chosen precise mathematical structure. This is not the case for the hypothesized genomic error-correcting codes

---

[1]An additional advantage is the availability of preexisting mathematical tools. As a striking example, Evariste Galois's theory of finite fields, long believed to exclusively belong to pure mathematics, found important engineering applications for designing error-correcting codes 120 years or so after being formulated. The vast family of algebraic codes rooted in Galois's work has not been dealt with in Ch. 6, but merely alluded to in Sec. 6.8.

where the alphabets themselves and their possible mathematical structure have to be determined. We put here 'alphabets' to the plural since we consider nested soft codes, and we already noticed that the component codes of a nested system may use different alphabets.

At first sight, however, it may seem obvious that the DNA alphabet, as consisting of the set of nucleotides $\{A, T, G, C\}$, should be considered as quaternary. But with what mathematical structure is it endowed? Liebovitch et al. [53] assumed it to be the ring of integers modulo 4. This choice is arbitrary and the structure considered in the coding literature for properly defining a linear code is that of a Galois field. It is only when the alphabet size $q$ is a prime that the addition rule modulo $q$ and that of the Galois field are identical, which is not the case for $q = 4$.

Even a mathematical structure more appropriate in engineering terms is unlikely to be relevant as being more or less arbitrarily chosen. Since soft codes can be defined by physical and chemical constraints, alphabets having a physical-chemical significance are more likely to be relevant. In this respect, we may consider that any nucleotide simultaneously belongs to two independent binary alphabets. First, the alphabet $\{R,Y\}$ whose symbols are the possible chemical structures of the nucleic bases, namely purine (2-cycle molecule, $A$ or $G$), denoted by $R$, or pyrimidine (single-cycle molecule, $T$ or $C$), denoted by $Y$. Second, the alphabet which we denote by $\{W,S\}$ where $W$, standing for 'weak,' represents the complementary pair of nucleotides $A$–$T$ which are tied together by two hydrogen bonds, and $S$, standing for 'strong,' the other complementary pair, namely $G$–$C$, where the nucleotides are tied together by three hydrogen bonds. The alphabet $\{R,Y\}$ corresponds to nucleic bases of different chemical structure and physical size, while the second one, $\{W,S\}$, indicates how strongly a nucleic base is tied with the complementary one. Then, Forsdyke interpreted a sequence of quaternary symbols as simultaneously bearing two independent binary codes, one over the alphabet $\{R,Y\}$ and the other one over $\{W,S\}$ [35]. According to the second Chargaff parity rule, the first code is balanced, i.e., the two symbols $R$ and $Y$ have the same frequency, like almost all codes designed by human engineers. On the contrary, the code over the alphabet $\{W,S\}$ is not balanced since the frequency of its symbols (referred to as the $G$+$C$ content) varies from a species to another one and, for long and inhomogeneous genomes like the human one, from a region to another one inside the genome. It could be interpreted as a kind of density modulation which maybe is read at several scales. The different number of hydrogen bonds of the two symbols of the alphabet $\{W,S\}$ implies that this density modulation results in a variation of the bonding energy between the complementary strands of DNA.

Other constraints are directly expressed in terms of larger alphabets. For instance, constraints induced on DNA by the structural properties of the proteins for which it 'codes' involve triplets of nucleic bases, i.e., the codons of the genetic 'code'. Then the alphabet size is $4^3 = 64$. Genes themselves may even be considered as the symbols of an alphabet [49, 61].

## 10.2.3 POTENTIAL GENOMIC SOFT CODES

We now list some potential genomic soft codes. Potential soft codes can first be associated with structural constraints of DNA. The very dense packing of DNA implies that any part of the DNA string is spatially close to many distant other parts of the same string (where 'distant' refers here to distance along the DNA string) [62]. The spatial neighborhood of these parts is due to chemical affinities[2], hence constraints, and many such constraints take place everywhere in the DNA string. The distances between the interactive parts along the DNA string may then play the role of interleavers so the whole genomic code would look like a kind of turbocode. Not only chemical affinities of the DNA string with distant parts of itself but also with histones, as constituents of nucleosomes, need be considered in the case of eukaryotic cells. The constraints due to the packing in nucleosomes concerns the 165 or so base pairs which are wrapped around histone octamers acting as spools, and the higher structures result in constraints presumably weaker but covering a wider range.

In the portions of the genome which specify proteins, i.e., in genes in a restricted sense, the sequence of codons (triplets of nucleotides) is furthermore constrained as are the proteins themselves: the structural constraints of proteins induce constraints on the sequence of codons which correspond to the amino-acids according to the 'genetic code'[3]. Physiologically active proteins are not fully characterized by the sequence of amino-acids (the polypeptidic chain) that the sequence of codons specifies. They owe their functional properties to their folding according to a unique pattern, which implies many chemical bonds between amino-acids which are separated along the polypeptidic chain but close together in the three-dimensional space when the protein is properly folded. Proteins are made of a number of three-dimensional substructures, especially $\alpha$-helices and $\beta$-sheets, which are themselves included into higher-order structures named 'domains,' which impose strong constraints of steric and chemical character on the succession of amino-acids. The sequence of amino-acids is thus subjected to many constraints. The sequence of codons which specify the amino-acids is thus itself subjected to related constraints. Due to the universal role of DNA for specifying proteins, such constraints must be present in any living being.

Soft Codes from Linguistic Constraints

We mentioned above that soft codes may be induced by linguistic constraints, too. We already noticed that the message which is needed for unambiguously identifying a biological species and even an individual inside it is very much shorter than the actual genomes, even the shortest ones like those of viruses (see Sec. 2.4). Genomes are thus highly redundant, a necessary condition for them to possess significant error-correcting properties. From another point of view, this redundancy has rather obvious reasons: the genome does not merely identify a living being. Modern biology interprets it as a *blueprint* for its construction. The genome of any living being needs actually contain the *recipe* for its development and its maintenance. Besides those parts of the genome which direct the synthesis of proteins, i.e., the genes in a restricted sense, and the associated regulatory sequences which switch on or off their expression (i.e., make the gene direct or not the synthesis of the protein it specifies), the genome must somehow *describe* the succession of operations which results in the development and the maintenance of a living thing. This demands some kind of *language*,

---

[2]'Chemical affinities' should be intended in a broad sense, including the forces which determine molecular shapes in an aqueous medium, e.g., the interaction of hydrophobic radicals with water, besides the usual chemical bounds.

[3]We use quotes here in order to remind that it is not truly a code in the information-theoretic sense, but rather a mapping in the mathematical vocabulary.

and a language involves many morphological and syntactic constraints which may be interpreted as generating redundant soft codes having error-correcting capabilities. Moreover, the linguistic constraints appear at several different levels, so a written text assumes the structure of nested soft codes which we were led to hypothesize for the genetic message. In order to illustrate the error-correcting ability of languages, notice that we, humans, can correctly perceive the spoken language in extremely noisy acoustic surroundings like vehicles or crowded places. By 'correctly perceive,' we do not mean to grasp the meaning, which concerns the semantic level, but simply to identify without error the uttered sequence of phonemes. It turns out that the individual phonemes are identified with a large error probability, but the linguistic constraints together with the high processing power of the human brain eventually result in errorless communication even in the presence of noise. We may assert that the daily experience of a conversation experimentally proves the ability of the human language, as a highly redundant soft code, to behave like good error-correcting codes designed by engineers. Biologists consistently use the metaphor of a written text to explain the role of the genetic message, at least in popular science books like [30] and many others. This metaphor is quite convincing, but the consequences it can have on the genome conservation are overlooked: indeed, it turns out that a language involves numerous and diverse constraints which may be interpreted as soft codes having error-correcting capabilities. Of course, it remains to understand how these error-correcting capabilities are exploited. Indeed, recent researches use tools of formal linguistics in order to describe the genomes and proteins [56, 71] but ignore the error correction problem. Moreover, the concept of dependence is actually shared by information and coding theory on the one hand, and formal linguistics applied to genomes and proteins on the other hand. As another interesting analogy, human natural languages have undergone evolution, and methods of the same kind as used for studying evolution in biology have been applied to that of languages.

The connection outlined in Sec. 5.5 between semantics and information implies that a longer genome is not only useful to decrease the error probability, but also provides room for 'more semantics' and thus enables specifying more complex beings. An important and useful tenet of information theory is the separation between information and semantics. However, the hypothesized error-correction mechanisms based on linguistic constraints heavily rely on the genome being a blueprint for the construction and maintenance of a living being, so one could consider the error-correction ability of the genetic message as, at least partially, a by-product of its semantics. But this is only a facet of the question. One can argue equally well that the error-correction property is the main feature of this message, since without it no transmission of hereditary characters would be possible and life could not have developed. Then, the construction and maintenance of living things would be a mere projection in the physical-chemical world of the abstract properties of the genetic message which enable error correction. This is a hen-and-egg problem, as often met in biology.

# 10.3 BIOLOGICAL SOFT CODES FORM NESTED SYSTEMS

Defining genomic codes by a set of constraints leads to a structure close to that described in Ch. 9.1 as a 'nested system,' although understandably much more complicated. Let us take as an example a eukaryotic gene. Consider that part of an exon which specifies the amino-acid sequence of an $\alpha$-helix in some protein. Certain amino-acids are preferred or forbidden in $\alpha$-helices. According to [24], Ala, Glu, Leu, and Met will thus be more often met than average in the considered sequence, while Pro, Gly, Tyr, and Ser will be almost absent. As a consequence, the codons **GC\*** for Ala, **GAA** and **GAG** for Glu, **UUA**, **UUG** and **CU\*** for Leu, and **AUG** for Met, will be more often met than average in the corresponding sequence of the messenger RNA, while the codons **CC\*** for Pro, **GG\*** for Gly, **UAU** and **UAC** for Tyr, and **UC\***, **AGU**, and **AGC** for Ser, will be almost absent (\* stands for any nucleotide **A, U, G,** or **C**; see Ch. 4). The DNA composition will thus greatly differ from the average in the part of the gene which corresponds to an $\alpha$-helix. Moreover, there are probably stronger constraints which forbid some successions of amino-acids as not compatible with the $\alpha$-helix structure. Similar constraints are related to other protein structures like $\beta$-sheets. A protein, moreover, involves at a larger scale several $\alpha$-helices and $\beta$-sheets organized in domains, according to a scheme which may be considered as a 'metacode' the symbols of which are words of the previously identified soft codes.

But these soft codes only pertain to the exons which actually direct the protein synthesis. According to a rather plausible conjecture (see Sec. 10.5) the nucleotides which belong to an exon are the information symbols of some block code in systematic form, the redundancy symbols of which are located in the corresponding introns. The previously identified soft codes are embedded into this block code which thus constitutes another coding layer of the nested system.

Only a fraction of the DNA base-pairs actually belongs to proper genes. This fraction is greatly variable from a species to another, and is quite small in, e.g., higher animals. We suggested above that the remainder, besides regulatory sequences, is encoded in linguistic codes necessary to describe the phenotype, which moreover assume themselves a nested structure.

This example shows the multiplicity of potential soft codes and their organization into a nested system or even into several intricate nested systems. The number and complexity of these soft codes is due in part to the choice of a eukaryotic gene as example. Both the block code made of exons and introns and the soft code due to the constraints of packing DNA into nucleosomes are not relevant to a prokaryotic gene. However, the close packing of DNA, although different, would similarly generate specific constraints. Those induced by the protein constraints would remain. The constraints of linguistic character are presumably still present, although less important since the specified phenotype is much simpler.

The number and variety of constraints indeed suggest that many potential genomic error-correcting mechanisms actually exist, which moreover are organized as nested soft codes. It is interesting to compare the resulting nested system with Barbieri's organic codes [5]. The codes we consider protect the DNA molecule against radiations and quantum indeterminism which no phenotypic shielding can ensure, hence are needed to ensure the conservation of hereditary information. Barbieri's concept of organic codes, on the other hand, does not refer to the necessity of error correction but results from a deep reflection on biological facts. He considers as an organic code the correspondence which exists between unidimensional[4] strings of completely different molecules (as a famous example, think of the relationship between triplets of nucleotides and the 20 amino-acids which make up proteins, referred to as the 'genetic code'). This correspondence does not result from any physical or chemical law, but should be considered as a pure convention or artifact, just like conventional rules are found in linguistic or engineering. (We conjectured that such rules are maintained thanks to 'semantic feedback loops' [17].) According to our point of view, the specific constraints obeyed by each of the strings which are tied together by a correspondence rule induce specific constraints in the DNA string, hence act as soft codes with error-correcting ability. Barbieri's organic codes actually assume the structure of nested codes. Especially significant in this respect is Fig. 8.2 in [5], p. 233, to be compared with Fig. 9.2 above which uses the fortress metaphor to illustrate the concept of nested codes. Moreover, both Barbieri's organic codes and the soft codes which make up our nested system successively appeared in geological times and correspond to major transitions in life evolution. This rather unexpected convergence provides a mutual confirmation of both points of view, which appear as complementary. This may also be thought of as an illustration of 'tinkering' as a typical method of nature, where some biological objects are used to perform functions completely different from those they initially performed. However, since many biological problems assume the chicken-and-egg form, a word like 'multivalence' could be more appropriate than 'tinkering' (although less picturesque) in the absence of a clear knowledge of the chronology.

Another point must be emphasized. The nested system provides a very unequal protection of different symbols depending on their place in the DNA string. Some base pairs are related to many others according to many constraints, and thus are strongly protected against casual errors. In our fortress metaphor, they belong to central coding layers. Others are subjected to few constraints and only related to few other base pairs and belong to outer layers. Some of the base pairs may even escape any constraint and appear as uncoded, belonging to the peripheral layer. Genomes, especially those of higher-living beings, thus appear as very inhomogeneous. This consequence of our hypotheses meets the experimental reality.

## 10.4 FURTHER COMMENTS ABOUT GENOMIC SOFT CODES

Soft codes do not exactly fit the properties of error-correcting codes which were described in Ch. 6. Since their definition involves probabilistic constraints and constraints expressed as inequalities, the mutual distances between their words become random, and especially the minimum distance $d$ which accounts to a large extent for the performance of a code. On the other hand, when discussing in Ch. 11 the consequences of our hypotheses we'll assume that the properties of genomic error-correcting codes are those of conventional codes. This may be thought of as a simplifying assumption. One may moreover argue that, if the soft codes combined into the nested scheme are numerous enough, and

---

[4]Unidimensionality is a common feature of messages in engineering and in biology. It appears as necessary for unambiguous semantic communication as well as in engineering.

if moreover their words are long enough, then the law of large number results in a small-variance minimum distance which can rightfully be approximated by a deterministic quantity.

Let us also notice that the soft code concept implies that the biological constraints are also those which enable error correction, at variance with the uncoded case but also with that of hypothetic codes obeying purely mathematical constraints. This may mean that the genomes which are admissible as words of a genomic error-correcting code also specify viable phenotypes. If this is true, decoding (regeneration) errors produce viable, possibly hopeful, monsters. This remark makes rather plausible the explanation of the Cambrian explosion to be suggested in Sec. 11.3.2.

## 10.5   IS A EUKARYOTIC GENE INVOLVING EXONS AND INTRONS A CODEWORD IN SYSTEMATIC FORM?

The interesting idea that introns are made of redundancy symbols associated with the message borne by the exons was formulated by Forsdyke as early as 1981 [34]. The literature generally states that introns are more variable than exons. A counter-example was however provided in 1995, again by Forsdyke, who experimentally found that the exons are more variable than introns in genes which 'code' for snake venoms [36].

It turns out that both the generally observed greater variability of introns and Forsdyke's counter-example can be explained by the assumption that the system of exons and introns actually acts as a systematic error-correcting code where exons constitute the information message (which directs the synthesis of a protein) and introns are made of the associated redundancy symbols[5]. Interpreted as a decoding error, a mutation occurs with large probability in favour of a codeword at a Hamming distance from the original word equal to the minimum distance of the code or slightly larger. If the exons 'code' for a protein of physiological importance, which is by far the most usual case, it may be expected that only mutations with a few errors within the exons, hence having no or little incidence on the protein, will survive natural selection. Few symbol errors being located in the exons, most of them will affect the introns since the total number of symbol errors after decoding is at least equal to the minimum distance of the code.

The situation is completely different in the case of genes which 'code' for snake venoms. The typical preys of snakes are rodents. Snakes and rodents are involved in an 'arms race': some rodents incur mutations which provide an immunity to snake venom, the population of rodents with such mutations increases as they escape their main predators, and the snakes are threatened with starvation unless mutations in their own genes make their venom able to kill mutated rodents [36]. The genes which 'code' for snake venoms are thus under high evolutive pressure: natural selection favours mutated genes producing proteins as different as possible from the original one. In terms of the Hamming distance, much of the difference should thus be located in the exons. The total number of decoding errors in exons and introns being roughly constant for a given code (equal to the minimum distance or slightly larger), introns are much less variable.

---

[5] An observer unaware of their role for error correction would easily conclude that the redundancy symbols have no function.

CHAPTER 11

# Biological Reality Conforms to the Main and Subsidiary Hypotheses

The simple model of Ch. 8, solely based on the main hypothesis, enabled calculating the lifetime, or *permanence*, of 'genomes' consisting of symbol sequences. Taking the subsidiary hypothesis into account as well as the results of Ch. 8 led in Ch. 9 to a qualitative description of a less simple, but more realistic, model of the living world. Introducing the concept of 'soft code' in Ch. 10 brought some biological relevance to that of genomic error-correcting code, and to that of 'nested system' central to the subsidiary hypothesis. (Remember that we used the word 'hypotheses' for lack of a more appropriate, stronger word, to designate inescapable conditions dictated by information theory.)

For translating properties of error-correcting codes into biological terms, we assume that the channel-encoded part of the genome (i.e., apart from the peripheral layer of the nested system, assumed to be left uncoded and to contain all individual differences, which constitutes only a small fraction of the genome) can be used in order to identify the species. Then, we may think of the information symbols of a genomic code as associated with the independent alternatives or parameters needed to specify a phenotype. The more numerous are the information symbols, i.e., the more numerous distinct dimensions has the genomic code, the more numerous semantic features a genome can instruct (see Sec. 5.5). The redundancy of genomes implies that the number of information symbols is only a small fraction of their total length. For a given redundancy rate, the phenotypic features that a genome determines are thus the more numerous, the larger its length.

We now show that the living world actually exhibits features which are biological consequences of these hypotheses. These are indeed the most basic and conspicuous properties of life, many of them being left essentially unexplained by conventional biology which rather deals with them as unquestioned postulates. Our hypotheses thus provide a broad understanding of the living world as a whole, which can be thought of as a global experimental check of their relevance. Moreover, they answer debated questions, some of them of very general character.

Table 11.1 lists in the left column a number of properties of error-correcting codes while the right column contains the features of the living world which result from each of these properties. We'll examine in the following sections each line of this table.

Table 11.1: Consequences of the main and subsidiary hypotheses: properties of error-correcting codes entail main features of the living world

| Error-correcting codes | Biology |
| --- | --- |
| Redundancy is necessary | Genomes are very redundant |
| Distance between codewords | Existence of discrete species |
| Nested codes | Hierarchical taxonomy |
| Limited correcting ability of a code | Necessity of successive regenerations |
| Regeneration errors result in distant codewords | Evolution proceeds by jumps |
| Nested codes imply a hierarchy of distances | Gradualism in periphery, saltationism in deep coding layers |
| The longer a code, the more efficient | Evolutive advantage of long genomes |
| Regeneration errors are chance events | Evolution is contingent |
| Close words do not correspond to close information messages | Close genomes do not correspond to close phenotypes (overlooked?) |

## 11.1   GENOMES ARE VERY REDUNDANT

The main hypothesis implies that genomes are redundant, in the sense that the sequences which belong to an error-correcting code, hence obey its constraints, are less than all the possible sequences of same length written using the same alphabet. We have already shown in Sec. 2.4 that genomes are extremely redundant. Indeed, the existing genomes are only a tiny fraction of all possible nucleotide sequences. Nobody knows the number of species which now live on Earth, because the identified ones are an unknown fraction of all extant species. An order of magnitude could be $10^7$. Let us assume that the extant species are a fraction of only 1/100 of all those which once existed (due to the extinction of many). We obtain $10^9$ as an order of magnitude for the number of past and present species. Although this rough estimate may look large (and should even this number be underestimated by a few orders of magnitude), it appears as extremely small with respect to the total number of possible genomes in the absence of redundancy, namely $4^n$. Typical values of the genome length $n$ range from thousands to billions and more base pairs, hence $4^n$ is a huge number. Genomes thus exhibit a very high redundancy.

## 11.2  LIVING BEINGS BELONG TO DISCRETE SPECIES

### 11.2.1  A GENOMIC ERROR-CORRECTING CODE IMPLIES DISCRETE SPECIES

The description of an error-correcting code we gave in Sec. 6.5 and the identification of a species with a genome (except for its small uncoded fraction) entails that the main hypothesis implies that *species are sparse.* It was already apparent in Ch. 8 that species discreteness is an essential feature of a genomic error-correcting code. We saw in Sec. 8.4 that the effects of natural selection could formally be taken into account in the toy living world which uses an error-correcting code, and we criticized the claim made by neo-Darwinians that the discreteness of species is the sole consequence of natural selection acting on phenotypes. We propose instead that this discreteness is *intrinsic* to the genomic error-correction system, hence pre-exists natural selection. Taxonomy then just reflects the discreteness of species, which itself follows from the necessary separation of codewords in the Hamming space of genomic error-correcting codes.

### 11.2.2  SPECIES CAN BE ORDERED ACCORDING TO A HIERARCHICAL TAXONOMY

Implying a nested error-correction system, the subsidiary hypothesis greatly complicates the picture, making it at the same time much more realistic. Instead of just predicting that species differ from each other, it entails a hierarchical taxonomy where species are the more different, they belong to deeper layers of the nested system. The subsidiary hypothesis then explains the existence of a hierarchical taxonomy, a major biological fact, which, moreover, necessarily coincides with the phyletic tree. Instead of the rather simple picture of a phyletic graph as introduced in Sec. 8.6, we find branchings at different hierarchical levels. The species become the more different, the deeper their phyletic trees diverge.

Each symbol of the genome belongs to some layer of the nested system. Considering these layers separately, we observe that going from the periphery to the center also means passing from a component code having very many codewords with a fairly high probability of gradual transitions between them to a component code having comparatively few but very different codewords with an extremely low probability of being transformed into each other.

### 11.2.3  TAXONOMY AND PHYLOGENY

We assume that the different coding layers appeared successively during geological times, meaning that the nested structure of ancestral forms of life has a number of layers less than that of more recent ones. As a consequence, the individuals of more recent species have both more diversified genomes and a longer time interval between regenerations (a larger lifetime for higher animals) than earlier species. The more numerous are the layers in the 'nested system' scheme, the larger the minimum distance in the innermost layers. The genome being made more resilient to errors, its conservation can be ensured with a larger time interval between regenerations (which has the advantage of more flexibility as regards environmental changes). Hence, we may expect that the longevity of individuals is larger in species where the genome involves many code layers. Similarly, in the same case, more differences between the individuals can exist since they are assumed to differ from each other only in the peripheral layer of the nested system, which is left uncoded. The specific features (i.e., common to all individuals of the species) are encoded into the most central layers where they are adequately protected. These remarks seem to roughly match reality if we think of the number of code layers as an index of how high a species is in the scale of evolution: compare, for instance, the complexity, lifetime, and individual diversity of *Caenorhabditis elegans* with those of vertebrates.

The appearance of the successive coding layers may well, at least for certain of them, coincide with the 'major transitions in evolution' [60] or with the onset of Barbieri's 'organic codes' [5]. The nested system indeed closely resembles Barbieri's organic codes, which, however, were not initially thought of as error-correcting: compare Fig. 9.2 with Fig. 8.3 of [5]. The many organic codes Barbieri convincingly identified may be considered as providing an experimental evidence of the existence of many codes organized according to the nested scheme. Each of them involves specific constraints so, according to the concept of soft code introduced in Ch. 10, they are potential error-correcting codes. The existence of many organic codes also suggests that the sketchy list of possible soft codes in the nested system given in Ch. 9 should be enriched with those associated with other organic codes in Barbieri's meaning.

The onset of the nested structure can be understood since DNA can be replicated. If its copy is appended to the initial genome instead of being separated from it, then an increase in the genome length results and the longer genome can evolve so as to increase its error-correcting ability. Using old means to create new ones, as assumed here, is a typical feature of nature's approach often referred to as *tinkering*. Moreover, many other mechanisms like inverse transcription or horizontal genomic transfer are already known as means for increasing the genome length.

## 11.3   NECESSITY OF SUCCESSIVE REGENERATIONS

### 11.3.1   CORRECTING ABILITY OF GENOMIC CODES AND TIME INTERVAL BETWEEN REGENERATIONS AS TARGETS OF NATURAL SELECTION

In the absence of genomic error-correcting code, it has been shown in Ch. 8 that the lifetime of a genome is the less, the larger is its length. Besides the decreasing permanence of individual genomes, it has been shown that no population of identical genomes, i.e., no species, can exist beyond a comparatively short genome length, of about $\ln(2)/p_{su}$ according to Eq. (8.2), where $p_{su}$ is the probability that an erroneous nucleotide is substituted for the correct one. For plausible estimates of $p_{su}$, this limit is much shorter than the actual genome length of many species, some of which being among the most enduring ones. Using for $p_{su}$ the estimate computed in Sec. 7.3 results in an upper limit of $3.5 \times 10^7$ base pairs. Many longer genomes exist in the actual living world, and there is no upper limit to the genome length of stable species: some of the oldest species have very long genomes. For instance, the lungfish genome contains about $1.4 \times 10^{11}$ base pairs, hence is 4,000 times larger than the above estimate of the limit. The absence of error-correction coding would forbid long genomes. The actual existence of species with long genomes confirms that genomes need error correction.

If, on the contrary, a genomic error-correcting code exists, the genome permanence, as computed in Sec. 8.2, varies approximately as $1/P_{cod}$, where $P_{cod}$ is the probability of a regeneration error. The genome permanence thus increases without limit if a long and efficient enough error-correcting code is used, provided the condition stated by the fundamental theorem of channel coding that the source entropy is less than the channel capacity is satisfied (see Sec. 5.3.1). Then $P_{cod}$ approaches 0 and the genome permanence has no upper limit.

The probability of a regeneration error, $P_{cod}$, depends itself on two factors: the intrinsic error-correcting ability of the code, on the one hand, and the time interval between regenerations $\Delta t$, on the other hand. The symbol error probability after a time interval $\Delta t$ is related to the substitution frequency $\nu_{su}$, assumed to be constant, according to Eq. (7.3). For quaternary symbols it reads:

$$p_{su}(\Delta\,(t)) = \frac{3}{4}\left[1 - \exp\left(-\frac{4}{3}\nu_{su}\Delta t\right)\right].$$

Then $p_{su}$ is an increasing function of $\Delta t$. The regeneration error probability, $P_{cod}$, is itself an increasing function of $p_{su}(\Delta t)$ which depends on the error-correcting code, and especially on its minimum distance $d$. If $p_{su}(\Delta t)$ is smaller than some threshold close to $d/2n$, $P_{cod}$ is very small; however, there is a steep increase of $P_{cod}$ when $p_{su}(\Delta t)$ approaches this threshold (see Sec. 6.5). The time interval $\Delta t$ is thus a critical parameter which controls the probability of successful regeneration. Too large its value makes the genomic error-correcting code inefficient. As directly controlling the genome permanence, the genomic error-correcting code and the time interval $\Delta t$ are the first and main targets of natural selection, independently of the associated phenotype.

## 11.3.2  NATURE MUST PROCEED WITH SUCCESSIVE REGENERATIONS

Not only genomic error-correcting codes must exist, but a genome must be regenerated before the number of cumulated errors exceeds the error-correcting ability of the code. To keep the number of erroneous symbols small enough, the time interval $\Delta t$ between regenerations should be short enough, or else regeneration errors result in the genome quickly tending to a sequence of nucleotides widely different from the initial codeword and almost uncorrelated with it (see Sec. 8.3). That nature proceeds with successive generations is a trivial biological fact. As a direct consequence of the hypothesis that a genomic error-correcting code exists, we find that genome conservation actually demands that successive **re**generations be performed, a much stronger statement. We moreover assume that the regeneration process concerns the genome as a whole so the several component codes of the nested system are not regenerated separately: the time interval $\Delta t$ between regenerations is shared by all these component codes.

Since conservation of a genome depends on its error-correcting ability on the one hand, and on the time interval between regenerations on the other hand, the proper matching of these presumably independent factors can only result from natural selection. Let us briefly look at consequences of their possible mismatch. If the time interval $\Delta t$ between regenerations is too large, the regeneration errors are rather frequent, giving rise to many comparatively short-lived species. The steep increase of the regeneration error probability when the nucleotide error rate approaches the threshold makes an even moderate increase of $\Delta t$ result in a large increase in mutation probability. Maybe the Cambrian explosion could be explained by too long an interval between regenerations soon corrected by natural selection ('soon' should be understood at the geological timescale, of course). If, on the contrary, $\Delta t$ is too small, regeneration errors very seldom occur. Then the species genome is very faithfully conserved but does not but very unfrequently originate in other ones. This species lacks flexibility to cope with environmental changes and thus its phylum risks extinction.

## 11.3.3  JOINT IMPLEMENTATION OF REPLICATION AND REGENERATION

We already noticed that regeneration and replication are conceptually different functions. Replication does not provide in itself any protection against the channel errors. It is its assumed error-correcting ability which enables regenerating the genome. However, a single individual is fragile and the survival of a species implies its replication besides regeneration. Both replication and regeneration of the genome must occur in living beings, as in the toy living world considered in Ch. 8. We now examine the possible modes of their joint implementation. Both can be performed simultaneously but there is no logical necessity for this. Clearly, once both functions of replication and regeneration began to be performed, the process of natural selection was engaged. What we observe is the joint product of regeneration, replication, and selection. Replication alone results in an exponentially increasing number of individuals, and the availability of finite resources necessarily results in selection.

The single cell of a unicellular being must be able to perform both replication and regeneration of its genome. The simplest way would thus be to perform genome regeneration systematically followed by replication at (roughly) periodic time intervals of $\Delta t$. It is, however, possible that the two events of regeneration and replication occur separately and are initiated by different means. For instance, replication may depend on an internal clock while regeneration-replication may be triggered by an external event.

In multicellular beings, each cell may behave as a unicellular being, hence be able to perform both regeneration and replication. This seems to be the case of, e.g., the Hydra. Again, replication may be performed apart from regeneration and be initiated by different means. There is, however, a new possibility, that of differentiated cells:

— somatic cells performing only replication in mitosis;

— germinal cells resulting from meiosis where regeneration is performed and followed with replication.

Due to the cost of regeneration (a much more complex operation than replication), differentiation into (comparatively) few germinal cells and many somatic cells has an evolutive advantage.

One may indeed plausibly assume that the genome regeneration occurs during meiosis and/or fertilization. Then some inner clock roughly determines the interval between them. But other mechanisms may be contemplated, especially triggered regeneration. For instance, the finding in *Arabidopsis thaliana* of 'non-Mendelian inheritance' [54] could be explained by assuming that, in this species and probably in other plants, the regeneration process is sporadically triggered by some external factor. Geneticists also suspect that some living beings (which are again plants) can control to some extent the mutation rate in response to external factors [48]. Triggering regeneration only under the control of external factors on the one hand, and favoring mutations to cope with the environmental situation, on the other hand, imply both that the mutation rate, i.e., the regeneration error rate, can be controlled, hence the existence of error-correcting means which can in certain circumstances be inhibited so as to increase the mutation rate. Notice however that increasing the mutation rate can also be obtained by inhibiting repair mechanisms which proofread the replicated genomes.

In the case of multicellular beings with differentiated germinal cells, a completely different way of performing regeneration also becomes possible: producing a large number of gametes using a simplified, hence approximate, regeneration algorithm and screening these gametes as regards their genome [70]. Screening must involve some criterion to distinguish the genomes which do not obey the coding constraints and should thus be discarded. Such a criterion implies testing whether a given nucleotide sequence belongs or not to the genomic code, which then works as *error detecting*. The complexity of error detection is generally much less than that of proper regeneration, i.e., determining what codeword is the closest to the received word. However, the full process involves growing a large population of germinal cells, hence a large number of replications. The high complexity of regenerating a single genome is replaced by many replications and many checks of faithfulness of replicated genomes to the genomic code.

The above remarks also suggest a possible connection with epigenetics. The regeneration process is complex, costs much energy, and is necessarily implemented by the phenotype. The crucial step of the regeneration process when the genomic message is rewritten can be perturbed by external conditions which affect the phenotype. The genome regeneration then depends to some extent on the environment, according to a weak kind of Lamarckian inheritance. This simple explanation avoids the recourse to hypothetical entities: DNA remains the one and only agent of heredity, but the dynamic mechanisms which ensure its conservation can be perturbed by external factors. From the experimental point of view, the study made by Marcus Pembrey et al. concerning inhabitants of the Swedish village of Överkalix reports evidence of transgenerational effects precisely related to well dated famine events and is especially interesting to us [63]. It turns out that the famine events had transgenerational consequences on the observed subjects only when they occurred during the meiosis from which the gametes of their grandparents originated. This is consistent with our hypothesis of a regeneration process which takes place at meiosis and can be affected by external factors. Maybe some phenomena referred to as epigenetic and explained by cytosine methylation or histone modification could find an alternative explanation in this remark.

## 11.4   SALTATIONISM IN EVOLUTION

### 11.4.1   REGENERATION ERRORS RESULT IN EVOLUTIVE JUMPS

The wide separation of genomes in terms of their Hamming distance which results from the main hypothesis, and the fact that regeneration errors, although rare events, occur with nonzero probability, entail that regeneration errors originate in new species, as illustrated by the model dealt with in Ch. 8. Insofar as it results from regeneration errors, biological evolution thus proceeds by jumps, i.e., is *saltationist*, although other mechanisms of speciation do exist. The subsidiary hypothesis however leads to complicate the picture as it makes the amplitude of a jump which results from a regeneration error strongly depend on the layer of the nested system where it takes place.

### 11.4.2   SALTATIONISM DEPENDS ON THE LAYER DEPTH IN THE NESTED SYSTEM

The subsidiary hypothesis that the genomic code is made of nested codes sheds light on its relationship with evolution. Let us consider separately the component codes of the nested system. To begin with, consider a long and very redundant code as found in the deeper layers (the words 'deep' and 'peripheral' refer to the fortress metaphor). Being highly redundant, this code has comparatively few words separated from each other by large Hamming distances, and we may assume that it is good as a result of natural selection. Then the probability of a regeneration error is very low, but its occurrence results in a very different genome, hence the corresponding phenotype is necessarily very different from that of the initial genome. In this layer, thus, strong saltationism occurs with very unfrequent transitions due to regeneration errors. In sharp contrast, we were already led to assume that the peripheral layer of the nested system is left uncoded. Then errors frequently occur but most often result in a genome very close to the original one. Close genomes will result in close phenotypes (apart from an important caveat stated in Sec. 11.7.2).

The peripheral layer of the nested system being left uncoded and the other layers being the more efficiently encoded, the deeper they are, we may think of the successive layers as exhibiting a transition between gradualism (in the peripheral layer) and saltationism, affecting more and more important features as deeper layers are considered. Remember that the protection against errors that the codes provide entirely relies on the distance between codewords. Their minimum distance is only $d = 1$ in the uncoded peripheral layer and increases from the periphery to the most central layers. The larger the distance between two codewords, the more they differ but the more improbable is a regeneration error transforming one of them into the other (see Fig. 9.3). In the nested system, thus, errors are frequent but result in small differences when they affect a symbol which belongs to the peripheral layer, while at the same time they become much less frequent and have more important consequences in deeper layers. Then microevolution which enables adaptation to small environmental changes depends on frequent errors in the peripheral layer, which is also the place where sexual reproduction introduces genetic diversity, while species differentiation and more important taxonomic divisions, i.e., macroevolution, result from very unfrequent but high-weight error patterns affecting deeper layers.

# 11.5  EVOLUTIVE ADVANTAGE OF LONG GENOMES; TREND OF EVOLUTION TOWARDS COMPLEXITY

## 11.5.1  EVOLUTIVE ADVANTAGE OF LONG GENOMES

The main hypothesis suffices to explain the poorly understood *trend of evolution towards increasing complexity*. What we mean is that new species or phyla more complex than the species or phyla already existing at the time of their appearance have an evolutive advantage. When dealing with the fundamental theorem of channel coding, we already noticed that the probability of a decoding (or regeneration) failure can be made the lower, the longer the words of the error-correcting code. This statement is paradoxical since increasing the word length also increases the average number of symbol errors in a received word. Appending a new layer in the nested system and provided the code is properly chosen, this theoretically proven property is fully confirmed by the engineering experience.

The genomic error-correcting code is thus presumably the more efficient, i.e., provides a lower probability of regeneration error $P_{cod}$, the longer it is. But we know from Ch. 8 that the permanence of a species is approximately proportional to its inverse, $1/P_{cod}$. Increasing the codeword length then provides an immediate evolutive advantage. The nested system postulated according to the subsidiary hypothesis, as described in Ch. 9, is precisely a means to both increase the codeword length and decrease the regeneration error probability (except in the peripheral layer).

## 11.5.2  INCREASING COMPLEXITY RESULTS FROM LENGTHENING GENOMES

Besides an immediate evolutive advantage, increasing the codeword length results in more numerous information symbols for a given redundancy rate. We already noticed in Sec. 5.5 that it gives room to more independent alternatives or parameters needed to specify phenotypic features. Even if it does not necessarily keep the redundancy rate constant, the onset of the nested system as described in Ch. 9 results in an increase in the number of information symbols every time a new coding layer is appended. More and more complex phenotypes can thus be specified as new coding layers are appended, and natural selection acting on these phenotypes lets the fittest ones survive.

# 11.6  EVOLUTION IS CONTINGENT

We saw in Ch. 8 that new genomes originate in regeneration errors although it is not the sole possible mechanism which creates new species. Regeneration errors are chance events which moreover occur at random instants, meaning that, to a large extent, biological evolution is contingent. We noticed moreover in Sec. 8.3 that the initial memory content of the genome is progressively forgotten as the number of regeneration errors increases.

Beware of not confusing a codeword chosen by such a chance event with a sequence of random symbols. As belonging to the genomic error-correcting code, the genome is actually subjected to many constraints which tie together its symbols (see Ch. 10). The wrong recovery of a codeword does not mean that each of its symbols is chosen at random. A regeneration error consists of choosing a wrong codeword *as a whole*. Errors on symbols which belong to a same codeword, at two specified locations, can by no means be considered as independent events and the probability of their simultaneous occurrence is larger than the product of the probabilities of these events taken separately. This suffices to refute arguments from the 'intelligent design' upholders on the improbability of errors simultaneously affecting two different nucleotides. Ironically, when biologists refute the 'intelligent design' argument, they implicitly rely on our main hypothesis, they generally ignore.

## 11.7   RELATIONSHIP BETWEEN GENOMES AND PHENOTYPES

### 11.7.1   GENOME AS SPECIFYING A PHENOTYPE

We may think of the information symbols of a genomic code as associated with the independent alternatives or parameters needed to specify a phenotype. The more numerous are the information symbols, i.e., the more numerous distinct dimensions has the genomic code, the more semantic features a genome can instruct (see Sec. 5.5). Remember that the redundancy of genomes implies that the information symbols constitute only a small fraction of the their total length. For a given redundancy rate, the phenotypic features that a genome determines are thus the more numerous, the larger its length.

In the nested system described in Ch. 9, the outer layers have a large dimension hence can represent many such phenotypic features. The inner layers in the nested system have however a dimension the smaller, the deeper they are. The most central ones can thus only represent few alternatives or parameters which then necessarily correspond to very basic phenotypic features, while the peripheral layers on the contrary correspond to more numerous but less basic features, in accordance with the taxonomic-phylogenic hierarchy.

### 11.7.2   NEIGHBORHOOD IN GENOMIC AND PHENOTYPIC SPACES

It is often considered as self evident that small differences between genomes result in small differences in the corresponding phenotypes. Deep mathematical reasons, however, prevent this to always be true. The space of genomes has an extremely large number of dimensions, equal to the genome length. The number of dimensions of the space of phenotypes, although large, is smaller since an error-correcting code, as assumed by the main hypothesis, is necessarily redundant. Then close neighbors in the space of genomes cannot always correspond to close neighbors in the space of phenotypes. Let us illustrate this property by an example where the numbers are likened to phenotypes and their decimal representations to genomes. (The base of the numeration system actually does not matter.) The space of numbers has a single dimension, in sharp contrast with the space of decimal representations which has as many dimensions as digits. Consider for instance the four 10-digit numbers $a = 1,999,999,999$, $b = 2,000,000,000$, $c = 2,000,000,001$, and $f = 9,000,000,000$. The first three are consecutive numbers, hence close to each other. However, while the Hamming distance between the decimal representations of $b$ and $c$ is 1 (the smallest possible distance between the representations of two different numbers), the distance between the decimal representations of $a$ and $b$ is 10, i.e., the largest possible between ten-digit numbers, although $a$ and $b$ are consecutive. The Hamming distance between the decimal representations of $b$ and $f$ also equals 1, although the numbers $b$ and $f$ widely differ. In order to map the numbers into their decimal representations, different weights need be affected to the digits so a difference in a single digit makes a big difference in the corresponding number if its weight is large.

### 11.7.3  ON GENOME COMPARISONS EXPRESSED AS PERCENTAGES

The previous section is a warning as concerns comparison of phenotypes which correspond to close genomes. Moreover, and independently of the above remark, the closeness of genomes is often expressed as a percentage of their length. For instance, the genome of the chimpanzee is often reported as differing from the human one by only 2%. But the human genome is more than 3 billion base pair long. The absolute difference, more than 60 million base pairs, is still huge and may correspond to a very large amount of information, hence of phenotypic difference.

CHAPTER 12

# Identification of Genomic Error-Correcting Codes, and other Open Problems

## 12.1 NECESSITY OF IDENTIFYING GENOMIC CODES

### 12.1.1 AN UNUSUAL APPROACH

Up until now, we asserted that (1) a genomic error-correcting code exists, (2) assumes the form of a nested system, (3) the components of which are soft codes. Statements (1) and (2), referred to above as our main and subsidiary hypotheses, express necessary conditions. Their necessity is of mathematical character, hence absolute: the relevance of information theory to literal communication is no more questionable than that of arithmetic or Euclidean geometry in practical situations. We saw in Ch. 11 that many actual basic features of the living world can be interpreted as mere consequences of these statements.

Statement (3) is not that necessary. It is just a plausible guess about what genomic error-correcting codes could look like. We saw in Ch. 10 that many potential soft codes *a priori* arise from biological constraints, and moreover that they assume the form of a nested system as described in Ch. 9. The following question, however, remains unanswered: *How are such constraints actually used in order to perform genome regeneration?* The answer would need experimentally identifying the biological regeneration mechanisms based on constraints.

The usual process of biological research starts with observations and experiments upon which theoretical constructions will possibly be based. Our approach in this work is inverse. Its starting point consists of a main and a subsidiary 'hypotheses' devoid of direct experimental bases but which are needed for conciliating two major but conflicting biological facts: the conservation of genomes and their modification by mutations. Information theory leaving no other alternative, genomic error-correcting codes *must* exist and the better conservation of very old genetic information *needs* that they take the form of a nested system.

If one wonders why such basic facts of life escaped as yet the investigations of biologists and remained unidentified, the answer is simple: they were not found because nobody searched for them. Those who could discover the solution ignored the problem. It will moreover become apparent that the experimental identification of genomic error-correcting codes is an extremely broad and difficult task. These codes must be deliberately searched for in the light of information theory.

### 12.1.2  A NECESSARY COLLABORATION OF ENGINEERS AND BIOLOGISTS

The experimental identification of the biological error-correction mechanisms, and especially of the means of genome regeneration, thus needs a close collaboration of biologists and information theorists. They should not only work together but actually learn from each others. There are many obstacles in this way; the cultural difference between mathematically educated people and experimental biologists is not the least one. Maybe the most blatant difficulty is the lack of an adequate popularization of information theory.

The very method to be used for identifying genomic error-correcting means has, moreover, to be invented. Even if the identification process is eventually successful, it will undoubtedly be long and uneasy. The progress will probably result from a slow mixing of the disciplines of information theory and genetics, and sudden breakthroughs are unlikely.

## 12.2   IDENTIFYING ERROR-CORRECTION MEANS

### 12.2.1  IDENTIFYING AN ERROR-CORRECTING CODE

Given a sequence of symbols, say, of nucleotides, consider the problem of determining whether this sequence is a word of some error-correcting code and, if so, of identifying the code to which it belongs. As stated, this problem has no solution. We defined an error-correcting code in Sec. 3.5 and Ch. 6 as a *collection* of words. For a hypothetical code defined by a list of words, as resulting from random coding, the knowledge of a particular codeword does not provide any clue on the other ones. The situation is a bit more favourable for a code defined by a set of constraints, but a single sequence can be characterized by an extremely large number of constraints. If it belongs to a code, the constraints specific to this code cannot be distinguished from many other possible ones. Identifying a code thus necessarily implies the knowledge of several of its codewords. In the case of a linear $(n, k)$ code (as defined in Sec. 6.4.2) at least $k$ words of the code must be available; $k$ words suffice provided they are linearly independent.

The problem resembles that of source coding of a given sequence without any knowledge of the source from which it originates. It is also, interestingly, almost the same problem as determining the algorithmic complexity of a sequence (see Sec. 5.4) by searching the smallest program which instructs a universal computer how to generate the given sequence. This problem has no general solution since the algorithmic complexity of a sequence is a proven uncomputable quantity.

Only collecting several different words can lead to identifying a code, with the further difficulty that the nested structure assumed according to the subsidiary hypothesis demands that the collected codewords belong to the same coding layer.

### 12.2.2  IDENTIFYING COMPONENT CODES OF THE NESTED SYSTEM

Identifying error-correction means would thus first imply acquiring an explicit knowledge of the several component codes of the nested system. Comparing genomes of closely related species could sometimes let know the constraints which characterize a code. For instance, the experimental comparison of different mutations affecting eukaryotic genes, some of which coding for snake venoms, can hopefully answer the question put in Sec. 10.5 and enable identifying this particular code for certain gene families.

Comparing genomes could first result in establishing a map of the component codes in the nested system. Some loci are much more variable than others, and some appear as almost invariant. A difficulty of this task is to determine the scale to be used. Should individual nucleotides be considered, or is a broader scale sufficient? Another point possibly worth studying would be to establish a correlation between the map of component codes and biological constraints. Establishing a map of the component codes in the nested system is clearly a very long process needing a great deal of experimental work. Even its completion would, however, be far from achieving the identification of genomic codes.

Studying species whose genomes notoriously resist damages due to radiation or drought, like *Deinococcus radiodurans* or Bdelloid rotifers, could possibly result in discovering especially efficient genomic error-correcting codes. The currently proposed explanations of their performance have recourse to templates, but no quantitative performance assessment is given for justifying them [81, 77, 41]. Indeed, the damages considered mainly consist of DNA 'backbone' breakings which cannot be repaired by merely using error-correcting codes, but it is unlikely that the events which break DNA do not also result in errors in the sequence of nucleotides, which need to be corrected.

### 12.2.3  IDENTIFYING REGENERATION MEANS

Even if the identification of certain codes is successful, the understanding of how they are used to regenerate parts of the genome is a most difficult, but mandatory, task. Indeed, an identified genomic error-correcting code has only a potential error-correcting ability. It is actually useful when, and only when, an algorithm implemented using the tools of molecular genetics is able to regenerate its words. The complexity of the decoding algorithms designed by engineers can easily be matched by processes of molecular genetics but the algorithms actually used for regenerating genomes are *a priori* unknown. Identifying how unknown algorithms are implemented is indeed a formidable task and the very method for doing so has to be invented. The ability of the ordinary tools of molecular genetics to perform elementary tasks of signal processing useful for genome regeneration should first be investigated. Many provisional assumptions followed by experiments intended to validate or disprove them will probably be necessary.

The genome regeneration process aims at producing a new genome subjected to all the constraints that a genome should obey. On the other hand, it should replicate the old genome which has possibly suffered errors. Since we likened the constraints of biological origin to soft codes, obeying constraints amounts to correct errors. Transposing in genetic terms the optimum regeneration rule we gave in Sec. 6.5.2, namely *choose the codeword the closest to the received word*, results in *choose the DNA sequence which obeys all the genomic constraints and is the closest to the genome to be regenerated*. We have seen in Sec. 6.6 that practical regeneration algorithms for convolutional codes actually operate symbol-by-symbol and moreover, in Sec. 6.7, that iterated soft input- soft output (SISO) decoding is especially efficient for regenerating turbocodes. The elementary mechanisms of template-replication already operate symbol-by-symbol. This suggests to look for more sophisticated, possibly iterated, regeneration mechanisms taking place during replication of prokaryotes or meiosis in sexed eukaryotes.

Among the constraints enumerated in Sec. 10.2.3, only some of them directly affect DNA strings. Other ones result from downstream constraints, for instance are induced by structural constraints which affect proteins. Still other constraints, the linguistic ones, have a more abstract character. It is easy to conceive that regeneration as defined takes account of the physical-chemical constraints directly affecting DNA. As regards the other ones, we must accept that they are in some way embodied in the molecular mechanism which performs regeneration. Identifying the regeneration means which specifically concern the DNA constraints then appears as a first, fairly easy step. Once this task is completed, that of identifying the other regeneration means will presumably be alleviated.

## 12.3   GENOME DISTINCTION AND CONSERVATION

The description of an error-correcting code as given in Ch. 6 shows that the possibility of distinguishing its elements, to be referred to for short as *distinction*, is its most important feature. A meaningful distance (Hamming's) has been introduced into the space of codewords, in terms of which the words which make up a code should be as separated from each other as possible. The set of distances between all pairs of words of a code is referred to as its *distance spectrum*. The channel errors which affect the transmitted word tend to blur the distinction between the received word and the codewords by diminishing their mutual distances. Decoding (or regeneration as we prefer to name it) consists of modifying the received word so as to restore the initial distance spectrum, hence the initial distinction (which may result in a regeneration error if too many channel errors occurred).

The analysis of sequences of nucleotides given by Forsdyke [37] may look entirely different, but almost all the reported genetic mechanisms result in *maintaining distinction*: for instance, insuring reproductive isolation between species or preserving gene integrity against recombination. Beyond the differences in vocabulary and points of view between separate disciplines, maintaining distinction may thus be thought of as a common basis of the theory of error-correcting codes and Forsdyke's evolutive bioinformatics. As an engineering discipline, the former is concerned with the problem of designing codes with an appropriate distance spectrum. The latter considers the discreteness of species and genes as prime biological facts and investigates the means which can maintain this discreteness. Both thus converge at the deepest level although their starting points are different. This level is that of *information theory*, which indeed may rightfully be referred to as the science of distinction.

Stated otherwise, the persistence of discreteness in the living world and the existence of genomic error-correcting means are two facets of one and the same thing. The reality of the former implies that of the latter. Discreteness is the price to be paid for conservation. Relevance of information theory as a basic background results, with the advantage of introducing quantitative methods and stating well defined limits, especially the channel capacity.

## 12.4 DIFFICULTIES OF CONCILIATING GENOMIC CODES AND SEXUAL REPRODUCTION

We were already led to conjecture that the peripheral layer of the nested system is left uncoded in order to make genomic encoding compatible with sexual reproduction. Then the genetic shuffling operated by crossing-over only affects the peripheral uncoded layer and the remainder of the genome is common to all members of a species.

Since regeneration errors result in genomes significantly different from the original ones, it is reasonable to think of these errors as a possible origin of species, hence that biological evolution is saltationist (as discussed in Sec. 11.4). On the one hand, it seems impossible to explain some biological facts unless assuming saltationism. To give a single example, Jean-Henri Fabre observed a female solitary wasp that stings a tarentula much bigger than her at a very precise point of its mouth so as to disable its venomous hooks [32]. She then paralyses the spider by stinging nervous ganglia and lays her egg inside it for feeding the larva after its hatching out. Were she not immediately successful in disabling the venomous hooks, she would be killed by the spider and her offspring would be lost. Can such an inherited behavior gradually appear? On the other hand, it is difficult to conciliate saltationism with sexual reproduction: how does Goldschmidt's 'hopeful monster' find a mate and produce viable offspring? This is an old objection to saltationism and the hypothesis that a genomic error-correcting code exists, which implies saltationism as an unavoidable consequence, does not make this objection stronger nor weaker. Too little is known as yet about the actual implementation of genomic error-correcting codes to address this problem.

Similar difficulties are met with 'horizontal genomic transfer' of any kind. This work has been exclusively devoted to the vertical transfer of genetic information, which is indeed essential in heredity. The compatibility of genomic error-correcting codes with horizontal genomic transfer is not obvious, although the onset of the nested system assumed to exist according to the subsidiary hypothesis necessarily relied on this kind of mechanisms.

CHAPTER 13

# Conclusion and Perspectives

Reviewing in [20] the book by Ed Regis, *What is Life? Investigating the Nature of Life in the Age of Synthetic Biology* [68], Steven Benner wrote:

> "Because building something requires a deep understanding of its parts, synthesis also stops scientists from fooling themselves. Data are rarely collected neutrally during analyses by researchers, who may discard some, believing the data to be wrong if they do not meet their expectations. Synthesis helps manage this problem. Failures in understanding mean that the synthesis fails, forcing discovery and paradigm change in ways that analysis does not."

Synthesis being the engineers' job, this remark is an excellent plea for a close collaboration of biologists and engineers. It is intended to genetic engineering but actually applies to any instance where nature and engineers are faced with the same problems.

Communication engineering benefits from the conceptual framework of information theory. Literal communication of symbol sequences ('literal' meaning that semantics is ignored) is actually a mathematical problem, and information theory is just that branch of mathematics which deals with it. On the other hand, heredity consists of literally communicating sequences of nucleotides over time, hence information theory is fully relevant to it.

Information theory can bring to genetics its concepts and methods as well as its results. Maybe its most important concept is that of *channel capacity*, a key notion which states an impassable limit to any communication. Information theory proves that communication without errors (more precisely, with an arbitrarily small error rate) is possible over a channel despite the errors which affect it, provided the information rate is less than the channel capacity, a quantity which decreases as the channel error rate increases. However, the very means which enable 'errorless' communication hinder any communication at all beyond the channel capacity. Both the possibility of 'errorless' communication below the capacity and its impossibility above it, although counterintuitive, are fully confirmed by the engineers' experience, besides being theoretically proven.

Applying the concept of channel capacity to genetics shows that the template-replication paradigm is wrong since merely copying genomes does not ensure their conservation. Genomes must act as words of an error-correcting code which enables their *regeneration*. Current genetics and evolution theory are deeply rooted in the implicit assumption that DNA strings, and especially genes, are permanent objects (which nevertheless suffer mutations). Drawing the consequences of their intrinsic impermanence demands that the bases of these sciences be drastically reinterpreted, a truly formidable task. Once information theory and error-correcting codes are recognized as relevant to biology, they impose to the living world severe constraints and limits which can no longer be ignored.

Finding that genomes, far from being intrinsically permanent objects, need a dynamic process to be conserved is not so surprising although it refutes a major tenet of current biology. Indeed, all other biological entities owe their conservation to dynamic processes, and we just state that genomes are no exceptions in this respect. One may rather wonder why they have been believed intrinsically permanent. That the results of science are provisional is a too often forgotten lesson of history. Good research generally provides questions as well as answers, and answering the new questions often leads to questioning the former answers. Considering that the present state of knowledge enables explaining once and for all how nature solves a problem is therefore an illusion. How heredity works has been wrongly believed for more than half a century as understood. The engineering point of view expounded in this book shows that it is actually a very tough problem. Its solution necessarily involves sophisticated engineering tools that most biologists ignore, namely, error-correcting codes. Finding a gap that wide in the current knowledge warns to be modest and cautious!

Revealing the fallacy of the template-replication paradigm is only the negative contribution of this work. On the positive side it opens perspectives for future research, some of them at most cursorily dealt with above. Among their possible benefits, let us mention improving the methods of biology, renewing the understanding of basic biological facts, and prompting discoveries.

As regards methodology, information theory makes the problems of communication met in biology accessible to computation. For instance, many mechanisms assumed to perform 'proof-reading' or 'DNA repair' based on the availability of two or more copies of the genomic message are described in the literature without any assessment of their expected performance. Once identified as solutions to engineering problems, assessing their performance becomes possible. Then the inadequacy of some of them would become obvious. Computation would act as regards biological hypotheses the same role as natural selection as regards species. Information theory provides means of such a performance assessment, and checking that a device or process actually fulfils a required performance is a current engineering practice which would be most beneficial to biology. Information theory can more generally provide a conceptual framework for genetics and even for biology as a whole. Its abstractness sharply contrasts with the current methods of biology but hopefully complements them by providing broad and general perspectives. The potential benefits to biology are worth the efforts necessary to assimilate information theory. Maybe the fusion of biology and information theory will eventually result in the long term in an extended biology.

As regards the renewed understanding of biological facts, it has been shown above that information-theoretic arguments explain features as basic to life as the necessity of successive generations and the existence of discrete species organized according to a hierarchical taxonomy, which moreover coincides with the phylogeny. They suggest that evolution proceeds by jumps and is basically contingent. Even the mysterious trend of evolution towards an increased complexity, which conventional biology fails to explain, appears as a mere consequence of natural selection operating on genomic error-correcting codes. Probably many other biological features can be similarly interpreted.

However, there is as yet no direct experimental evidence of the existence of genomic error-correcting codes. Revealing its necessity prompts investigating how such codes are implemented, and especially how the genome regeneration they enable is performed. The novelty of this approach entails that even its methods have to be invented. The vision provided by information theory turns the old living world into a new world to be explored.

# Bibliography

[1] B. Audit, C. Vaillant, A. Arneodo, Y. d'Aubenton-Carafa, and C. Thermes, "Long-range correlation between DNA bending sites: relation to the structure and dynamics of nucleosomes," *J. Mol. Biol.*, Vol. 316, 2002, pp. 903–918. DOI: 10.1006/jmbi.2001.5363

[2] O. Avery, M. McCarty, and C. MacLeod, "Studies of the chemical nature of the substance inducing the transformation of pneumococcal types. Induction of transformation by a desoxyribonucleic acid fraction isolated from Pneumococcus Type III," *J. Exp. Med.*, Vol. 79, 1944, pp. 137–158. DOI: 10.1084/jem.79.2.137

[3] L.R. Bahl, J. Cocke, F. Jelinek, and J. Raviv, "Optimal decoding of linear codes for minimizing symbol error rate," *IEEE Trans. Infor. Theor.*, Vol. IT-20, Mar. 1974, pp. 284–287. DOI: 10.1109/TIT.1974.1055186

[4] D. Baltimore, "Our genome unveiled," *Nature*, Vol. 409, No. 6822, Feb. 2001, pp. 814–816. DOI: 10.1038/35057267

[5] M. Barbieri, *The Organic Codes*, Cambridge University Press, Cambridge, 2003.

[6] G. Battail and M. Decouvelaere, "Décodage par répliques," *Annales Télécommunic.*, Vol. 31, No. 11-12, Nov.-Dec. 1976, pp. 387–404.

[7] G. Battail, M. Decouvelaere, and P. Godlewski, "Replication decoding," *IEEE Trans. Infor. Theor.*, Vol. IT-25, No. 3, May 1979, pp. 332–345. DOI: 10.1109/TIT.1979.1056035

[8] G. Battail, "Décodage pondéré optimal des codes linéaires en blocs I.- Emploi simplifié du diagramme du treillis," *Annales Télécommunic.*, Vol. 38, No. 11-12, Nov.-Dec. 1983, pp. 443–459.

[9] G. Battail, "Pondération des symboles décodés par l'algorithme de Viterbi," *Annales Télécommunic.*, Vol. 42, No. 1-2, Jan.-Feb. 1987, pp. 31–38.

[10] G. Battail, "Construction explicite de bons codes longs," *Annales Télécommunic.*, Vol. 44, No. 7-8, Jul.-Aug. 1989, pp. 392–404.

[11] G. Battail, "We can think of good codes, and even decode them," *Eurocode 92*, Udine, 26-30 Oct. 92. CISM courses and lectures No. 339, P. Camion, P. Charpin, and S. Harari, Eds., Springer, 1993, pp. 353–358.

[12] G. Battail, "On random-like codes," *Proc. 4-th Canadian Workshop on Information Theory*, Lac Delage, Québec, 28-31 May 1995. Lecture Notes in Computer Science No. 1133, Springer, 1996, pp. 76–94.

[13] G. Battail, "Does information theory explain biological evolution?", *Europhys. Lett.*, Vol. 40, No. 3, Nov. 1st, 1997, pp. 343–348. DOI: 10.1209/epl/i1997-00469-9

[14] G. Battail, "On Gallager's low-density parity-check codes," International Symposium on Information Theory, *Proc. ISIT 2000*, Sorrento, Italy, 25-30, juin 2000, p. 202, DOI: 10.1109/ISIT.2000.866500

[15] G. Battail, "Is biological evolution relevant to information theory and coding?", *Proc. ISCTA '01*, Ambleside, UK, Jul. 2001, pp. 343–351.

[16] G. Battail, "An engineer's view on genetic information and biological evolution," *Biosystems*, Vol. 76, 2004, pp. 279–290. DOI: 10.1016/j.biosystems.2004.05.029

[17] G. Battail, "Genetics as a communication process involving error-correcting codes," *J. Biosemiot.*, Vol. 1, No. 1, 2005, pp. 103–144.

[18] G. Battail, "Should genetics get an information-theoretic education?", *IEEE Eng. Med. Biol. Mag.*, Vol. 25, No. 1, Jan.-Feb. 2006, pp. 34–45. DOI: 10.1109/MEMB.2006.1578662

[19] G. Battail, "Information theory and error-correcting codes in genetics and biological evolution," in *Introduction to Biosemiotics*, Barbieri, M., Ed., Springer, 2006. DOI: 10.1007/1-4020-4814-9_13

[20] S. Benner, "Biology from the bottom up," *Nature*, Vol. 452, No. 7188, 10 avr. 2008, pp. 692–694, DOI: 10.1038/452692a

[21] C. Berrou, A. Glavieux, and P. Thitimajshima, "Near Shannon limit error-correcting coding and decoding: turbo-codes," *Proc. ICC'93*, Geneva, Switzerland, May 1993, pp. 1064–1070. DOI: 10.1109/ICC.1993.397441

[22] C. Berrou and A. Glavieux, "Near optimum error correcting coding and decoding: turbo codes," *IEEE Trans. Commun.*, Vol. 44, Oct. 1996, pp. 1261–1271, DOI: 10.1109/26.539767

[23] C. Berrou, "Métamorphoses et promesses du codage correcteur d'erreurs," Seminar organized for the election of Claude Berrou at the French Academy of Sciences, Jun. 2, 2008.

[24] C. Branden and J. Tooze, *Introduction to Protein Structure*, New York and London, Garland, 1991.

[25] G. Chaitin, *Metamaths!*, New York: Pantheon Books, 2005.

[26] E. Chargaff, "How genetics got a chemical education," *Ann. New York Acad. Sci.*, Vol. 325, 1979, pp. 345–360. DOI: 10.1111/j.1749-6632.1979.tb14144.x

[27] T.M. Cover and J.A. Thomas, *Elements of Information Theory*, New York: Wiley, 1991.

[28] P. David and S. Samadi, *La théorie de l'évolution*. Paris: Flammarion, 2000.

[29] R. Dawkins, *The Selfish Gene*, Oxford: Oxford University Press, 1976; new edition, 1989.

[30] R. Dawkins, *The Blind Watchmaker*, New York, Norton, Longman, 1986.

[31] P. Elias, "Coding for noisy channels," *IRE Conv. Rec., Part 4*, New York, 1955, pp. 37–46.

[32] J.-H. Fabre, *Souvenirs Entomologiques*, Paris: Delagrave, 1880.

[33] G.D. Forney, Jr, "The Viterbi algorithm," *Proc. IEEE*, Vol. 61, 1973, pp. 268–278. DOI: 10.1109/PROC.1973.9030

[34] D.R. Forsdyke, "Are introns in-series error-detecting sequences?", *J. Theor. Biol.*, Vol. 93, 1981, pp. 861–866. DOI: 10.1016/0022-5193(81)90344-1

[35] D.R. Forsdyke, http://post.queensu.ca/ forsdyke/

[36] D.R. Forsdyke, "Conservation of stem-loop potential in introns of snake venom phospholipase $A_2$ genes. An application of FORS-D analysis," *Mol. Biol. Evol.*, Vol. 12, 1995, pp. 1157–1165.

[37] D.R. Forsdyke, *Evolutionary Bioinformatics*, New York, Springer, 2006.

[38] R.E. Franklin and R.G. Gosling, "Molecular configuration in sodium thymonucleate," *Nature*, Vol. 171, No. 4356, 25 Apr. 1953. pp. 740–741. Reprinted in *Nature*, Vol. 421, No. 6921, Jan. 23, 2003, pp. 400–401.

[39] R.G. Gallager, "Low density parity-check codes," Cambridge, MA: MIT Press, 1963.

[40] R.G. Gallager, "A simple derivation of the coding theorem and some applications," *IEEE Trans. Infor. Theor.*, Vol. IT-13, No. 1, Jan. 1965, pp. 3–18. DOI: 10.1109/TIT.1965.1053730

[41] E. Gladyshev and M. Meselson, "Extraordinary resistance of Bdelloid rotifers to ionizing radiation," *Proc. Ntl. Acad. Sci. USA*, Vol. 105, 2008, pp. 5139–5144. DOI: 10.1073/pnas.0800966105

[42] M.J.E. Golay, "Notes on digital coding," *Proc. IRE*, Vol. 37, No. 6, Jun. 1949, p. 637.

[43] E. Guizzo, "Closing in on the perfect code," *IEEE Spectrum*, Vol. 41, No. 3 (INT), Mar. 2004, pp. 28–34. DOI: 10.1109/MSPEC.2004.1270546

[44] J. Hagenauer and P. Hoeher, "A Viterbi algorithm with soft-decision outputs and its applications," *Proc. GLOBECOM'89*, Nov. 1989, Dallas, pp. 1680–1686.

[45] R.W. Hamming, "Error detecting and error correcting codes," *Bell Syst. Tech. J.*, Vol. 29, No. 1, Jan. 1950, pp. 147–160.

[46] D.A. Huffman, "A method for the construction of minimum redundancy codes," *Proc. IRE*, Vol. 40, 1952, pp. 1098–1101. DOI: 10.1109/JRPROC.1952.273898

[47] International Human Genome Sequencing Consortium, "Initial sequencing and analysis of the human genome," *Nature*, No. 6822, Vol. 409, Feb. 15, 2001, pp. 860–921.

[48] E. Jablonka and M.J. Lamb, *Epigenetic Inheritance and Evolution. The Lamarckian Dimension*, Oxford: Oxford University Press, 1995.

[49] S.A. Kauffman, *The Origins of Order*, Oxford: Oxford University Press, 1993.

[50] A.I. Khinchin, *Mathematical Foundations of Information Theory*, New York, Dover 1957.

[51] A.N. Kolmogorov, "Logical basis for information theory and probability theory," *IEEE Trans. Infor. Theor.*, Vol. IT-14, No. 5, Sept. 1968, pp. 662–664. DOI: 10.1109/TIT.1968.1054210

[52] A.S. Kondrashov, "Direct estimate of human per nucleotide mutation rate at 20 loci causing Mendelian diseases," *Human Mutation*, Vol. 21, No. 1, Jan. 2003, pp. 12–27. DOI: 10.1002/humu.10147

[53] L.S. Liebovitch, Y. Tao, A.T. Todorov, and L. Levine, "Is there an Error Correcting Code in the Base Sequence in DNA?", *Biophys. J.*, Vol. 71, 1996, pp. 1539–1544.

[54] S.J. Lolle, J.L. Victor, J.M. Young, and R.E. Pruitt, "Genome-wide non-mendelian inheritance of extra-genomic information in *Arabidopsis*," *Nature*, Vol. 434, No. 7032, Mar. 24, 2005, pp. 505–509. DOI: 10.1038/nature03380

[55] F.J. MacWilliams and N.J.A. Sloane, *The Theory of Error-Correcting Codes*, Amsterdam: North Holland, 1977.

[56] R.N. Mantegna, S.V. Buldyrev, A.L. Goldberger, S. Havlin, C.-K. Peng, S. Simons, and H.E. Stanley, "Linguistic features of noncoding DNA sequences," *Phys. Rev. Lett.*, Vol. 73, 1994, pp. 3169–3172. DOI: 10.1103/PhysRevLett.73.3169

[57] J.L. Massey, *Threshold Decoding*, Cambridge, MA: MIT Press, 1963.

[58] J. Maynard Smith, "The theory of games and the evolution of animal conflicts," *J. Theorect. Biol.*, Vol. 47, 1974, pp. 209–221.

[59] J. Maynard Smith, "The idea of information in biology," *Quart. Rev. Biol.*, 74(4), 1999, 395–400.

[60] J. Maynard Smith and E. Szathmáry, *The Major Transitions in Evolution*, W.H. Freeman, 1995, New York, Oxford: Oxford University Press, 1997.

[61] O. Milenkovic, "The information processing mechanism of DNA and efficient DNA storage," DIMACS working group on theoretical advances in information recording, Mar. 22-24, 2004.

[62] B. Moher, "Spring theory," *Nature*, Vol. 448, No. 7157, Aug. 30, 2007, pp. 984–986. DOI: 10.1038/448984a

[63] M.E. Pembrey, L.O. Bygren, G. Kaasti, S. Edvinsson, K. Northstone, M. SJöström, and J. Golding. "Sex-specific, male-line transgenerational responses in humans," *Eur. J. Hum. Genet.*, Vol. 14, 2006, pp. 156–166. DOI: 10.1038/sj.ejhg.5201538

[64] M.W. Nachman, "Haldane and the first estimates of the human mutation rate," *J. Genet.*, Vol. 83, No. 3, Dec. 2004, pp. 231–233. DOI: 10.1007/BF02717891

[65] W.W. Peterson and E.J. Weldon, Jr., *Error-Correcting Codes*, 2nd ed., Cambridge, MA: MIT Press, 1972.

[66] J.R. Pierce, "The early days of information theory," *IEEE Trans. Infor. Theor.*, Jan. 1973, pp. 3–8. DOI: 10.1109/TIT.1973.1054955

[67] M. Radman, "Fidelity and infidelity," *Nature*, Vol. 413, No. 6852, Sept. 13, 2001, p. 115. DOI: 10.1038/35093178

[68] E. Regis, *What is Life? Investigating the Nature of Life in the Age of Synthetic Biology*, New York, Farrar, Straus and Giroux, 2008.

[69] M. Ridley, *Mendel's Demon: [Gene Justice and the Complexity of Life]*, London: Weidenfeld and Nicholson, 2000.

[70] M.E. Samuels, Private communication (Aug. 2007).

[71] D.B. Searls, "The language of genes," *Nature*, Vol. 420, No. 6912, Nov. 14, 2002, pp. 211–217. DOI: 10.1038/nature01255

[72] C.E. Shannon, "A mathematical theory of communication," *Bell Syst. Tech. J.*, Vol. 27, Jul. and Oct. 1948, pp. 379–457 and pp. 623–656.

[73] E. Szathmáry and J.M. Smith, "The major evolutionary transitions," in *Shaking the Tree: Readings from* Nature *in the History of Life*, H. Gee, Ed., 2000, pp. 32–47. The University of Chicago Press. DOI: 10.1038/374227a0

[74] A.J. Viterbi, "Convolutional codes and their performance in communication systems," *IEEE Trans. Commun. Tech.*, Vol. COM-19, Oct. 1971, pp. 751–772. DOI: 10.1109/TCOM.1971.1090700

[75] R.F. Voss, "Evolution of long-range fractal correlation and $1/f$ noise in DNA base sequences," *Phys. Rev. Lett.*, Vol. 68, Jun. 1992, pp. 3805–3808. DOI: 10.1103/PhysRevLett.68.3805

[76] J.D. Watson and F.H.C. Crick, "Molecular structure of nucleic acids," *Nature*, Vol. 171, No. 4356, Apr. 25, 1953, pp. 737–738. Reprinted in *Nature*, Vol. 421, No. 6921, Jan. 23, 2003, pp. 397–398.

[77] D.B. Mark Welch, J.L. Mark Welch, and M. Meselson, "Evidence for degenerate triploidy in Bdelloid retifers," *Proc. Ntl. Acad. Sci. USA*, Vol. 105, 2008, pp. 5145–5149.

[78] J.K. Wolf, "Efficient maximum likelihood decoding of linear block codes using a trellis," *IEEE Trans. Infor. Theor.*, Vol. IT-24, No. 1, Jan. 1978, pp. 76–80. DOI: 10.1109/TIT.1978.1055821

[79] H.P. Yockey, *Information Theory and Molecular Biology*, Cambridge: Cambridge University Press, 1992.

[80] H.P. Yockey, *Information Theory, Evolution, and the Origin of Life*, Cambridge: Cambridge University Press, 2005.

[81] K. Zahradka, D. Slade, A. Bailone, S. Sommer, D. Averbeck, M. Petranovic, A.B. Lindner, and M. Radman, "Reassembly of shattered chromosomes in D. radiodurans," *Nature*, Vol. 443, No. 7111, Oct. 5, 2006, pp. 569–573. DOI: 10.1038/nature05160

# Biography

**Gérard Battail** was born in Paris, France, on June 5, 1932. He graduated at the Faculté des Sciences (1954) and the Ecole nationale supérieure des Télécommunications (ENST) in 1956, both in Paris. After his military duty, he joined the Centre national d'Etudes des Télécommunications (CNET) in 1959. He worked there on modulation systems and especially on frequency modulation, using fundamental concepts of information theory to understand its behaviour in the presence of noise, especially the threshold effect. In 1966, he joined the Compagnie française Thomson-Houston (later become Thomson-CSF) as a scientific advisor to technical teams designing radioelectric devices. There he interpreted channel coding as a diversity system for designing decoders, especially soft-input ones. He also worked on source coding, frequency synthesizers, mobile communication and other problems related to the design of industrial radiocommunication devices. In 1973, he joined ENST as a Professor. He taught there modulation, information theory and coding. He had also research activities in the same fields with special emphasis on adaptive algorithms as regards source coding and, for channel coding, on soft-in, soft-output decoding of product and concatenated codes. He proposed as a criterion for designing good codes the closeness of its distance distribution with respect to that of random coding instead of maximizing the minimum distance. These rather unorthodox views are now recognized as having paved the way to the invention of turbocodes by Berrou and Glavieux in the early 90s. After his retirement in 1997, he started working on applications of information theory to the sciences of nature. He especially investigated the role of information theory and error-correcting codes in genetics and biological evolution, showing that the conservation of genomes needs error-correcting means.

He applied for many patents, wrote many papers and participated in many symposia and workshops. He also authored a textbook on information theory published by Masson in 1997. He is a member of the Société de l'Electricité, de l'Electronique, des Technologies de l'Information et de la Communication (SEE) and of the Institute of Electrical and Electronics Engineers (IEEE). Before his retirement, he was a member of the editorial board of the *Annales des Télécommunications*. From 1990 to 1997, he was the French official member of Commission C of URSI (International Radio-Scientific Union). From June 2001 to May 2004, he served as Associate Editor at Large of the *IEEE Transactions on Information Theory*.

# Index

Printed in the United States
by Baker & Taylor Publisher Services